Urban Sustainability

The Urban Sustainability Book Series is a valuable resource for sustainability and urban-related education and research. It offers an inter-disciplinary platform covering all four areas of practice, policy, education, research, and their nexus. The publications in this series are related to critical areas of sustainability, urban studies, planning, and urban geography.

This book series aims to put together cutting-edge research findings linked to the overarching field of urban sustainability. The scope and nature of the topic are broad and interdisciplinary and bring together various associated disciplines from sustainable development, environmental sciences, urbanism, etc. With many advanced research findings in the field, there is a need to put together various discussions and contributions on specific sustainability fields, covering a good range of topics on sustainable development, sustainable urbanism, and urban sustainability. Despite the broad range of issues, we note the importance of practical and policy-oriented directions, extending the literature and directions and pathways towards achieving urban sustainability.

The series will appeal to urbanists, geographers, planners, engineers, architects, governmental authorities, policymakers, researchers of all levels, and to all of those interested in a wide-ranging overview of urban sustainability and its associated fields. The series includes monographs and edited volumes, covering a range of topics under the urban sustainability topic, which can also be used for teaching materials.

Ali Cheshmehzangi · Jian Zuo · Ayyoob Sharifi ·
Rongpeng Zhang · Abbas Ziafati Bafarasat ·
Jie Zhao
Editors

Design Innovation for Climate-Resilient Cities and Built Environments

Architectural and Technological Solutions

Springer

Editors
Ali Cheshmehzangi
School of Architecture, Design and Planning
University of Queensland
Brisbane, QLD, Australia

Ayyoob Sharifi
The IDEC Institute
Hiroshima University
Hiroshima, Hiroshima, Japan

Abbas Ziafati Bafarasat
School of the Built Environment
Oxford Brookes University
Oxford, Oxfordshire, UK

Jian Zuo
School of Architecture and Civil Engineering
University of Adelaide
Adelaide, SA, Australia

Rongpeng Zhang
School of Architecture and Planning
Hunan University
Changsha, Hunan, China

Jie Zhao
Stuart Weitzman School of Design
University of Pennsylvania
Philadelphia, PA, USA

P3Global
Dubai, UAE

ISSN 2731-6483 ISSN 2731-6491 (electronic)
Urban Sustainability
ISBN 978-981-92-2296-4 ISBN 978-981-92-2297-1 (eBook)
https://doi.org/10.1007/978-981-92-2297-1

This Springer imprint is published by the registered company Springer Nature Singapore Pte Ltd.
The registered company address is: 152 Beach Road, #21-01/04 Gateway East, Singapore 189721, Singapore

Dedicated to those who believe that design innovation is not merely technical progress, but a shared commitment to resilient, just, and life-supporting cities and built environments.

Preface

> The first rule of sustainability is to align with natural forces, or at least not try to defy them.
>
> — *Paul Hawken, a renowned environmentalist, author, and activist*

Cities are humanity's most ambitious creations. They are engines of culture, trade, innovation, and collective aspiration. Yet they are also places where our contradictions become most visible. Towers rise where wetlands once absorbed floodwaters. Highways carve through landscapes shaped by centuries of ecological balance. Urban expansion promises progress while simultaneously exposing the fragile systems that sustain life. Nowhere are these tensions more evident than in the accelerating pressures of climate change.

Paul Hawken's straightforward statement encapsulates a basic reality: sustainability is neither a policy catchphrase nor a technical ploy. The natural processes that regulate the world and human ambition have a basic link. Cities become unstable when they disregard these systems and build against the wind, water, soil, and climate. Floods get worse. Heat builds up. Infrastructure breaks down. Communities become more susceptible. Society ultimately bears the price of disobeying nature. Thus, this book emerges from that recognition.

Urban regions around the world are about to enter a period characterized by environmental unpredictability. Prolonged droughts, severe storms, rising temperatures, and changing ecological patterns are no longer far-off forecasts. They are already changing how communities live their daily lives, how buildings operate, and how cities operate. Resilience is now a new requirement for the built environment, which was previously primarily intended for efficiency, expansion, and permanence.

A change of perspective is necessary for resilience. It challenges long-held beliefs about development, infrastructure, and design among planners, architects, engineers, legislators, and citizens. Can cities learn to cooperate with natural forces rather than oppose them? Can urban systems anticipate, adapt, and evolve instead of reacting to crises after they happen? Can integrated techniques change our perception of the built environment's future in place of disjointed solutions? These questions lie at the heart of this book.

The chapters included here examine the relationship between urban design, technological innovation, and climate resilience. They show a quickly changing terrain of concepts, instruments, and methods that are reshaping how cities might address environmental issues. The contributions in this volume show how research and practice are coming together to address some of the most important urban issues of our day, from digital twin technologies and smart building systems to green infrastructure and climate-responsive planning. Yet this book is not simply a collection of technical solutions. It is also an invitation to think differently about the role of knowledge, creativity, and collaboration in shaping urban futures. What if resilience were treated not as a defensive strategy, but as a driver of innovation? What if climate challenges became opportunities to rethink the form, function, and meaning of our cities? What if the built environment evolved from a system that consumes natural resources into one that regenerates them?

Such possibilities appear to be more than just theoretical, according to the investigations included in this collection. Researchers and practitioners worldwide are experimenting with novel strategies that integrate human-centered design, ecological thinking, and digital intelligence. They show how urban planning may reduce heat and flooding, how technological systems can improve efficiency and well-being, and how predictive modeling can foresee environmental problems.

These investigations also highlight the intricacy of change. Sustainable results cannot be ensured by technological innovation alone. The success or failure of resilience techniques is influenced by social acceptance, governance systems, economic realities, and cultural contexts. Therefore, creating climate-responsive cities calls for more than just cutting-edge equipment; it also calls for interdisciplinary discussion, institutional bravery, and long-term planning.

The built environment professions have a great deal of responsibility in this regard. Policymakers, engineers, planners, architects, and designers do more than just create physical environments. They are influencing how cultures will deal with the environmental issues of the upcoming decades. Their choices have an impact on how cities interact with the surrounding ecosystems, how infrastructure survives storms, and how communities respond to heatwaves.

Urban resilience's future won't be decided by a single field, piece of technology, or set of regulations. Rather, it will result from the seamless integration of innovation and tradition, data and experience, and science and design. It will rely on our abilities as a group to work in harmony with natural forces rather than in opposition to them. That alignment is not a constraint. It is an opportunity.

Cities may become more efficient, adaptable, and ultimately more compassionate places to live if they can learn to cooperate with ecological cycles, climatic systems, and natural processes. It will take creative imagination, critical thinking, and experimentation to find the way forward. It will also necessitate a readiness to consider how, why, and what type of urban future we really want. Readers are encouraged to participate in that investigation via this book.

Brisbane, Australia	Ali Cheshmehzangi
Adelaide, Australia	Jian Zuo
Hiroshima, Japan	Ayyoob Sharifi
Changsha, China	Rongpeng Zhang
Oxford, UK	Abbas Ziafati Bafarasat
Philadelphia, USA	Jie Zhao

Acknowledgements

We gratefully acknowledge the collective contributions that have shaped this book. We extend our sincere appreciation to all authors, co-authors, chapter contributors, and collaborators whose expertise, commitment, and creativity have advanced the discourse on climate-resilient cities and built environments. We also recognize the strong foundation laid by our earlier collaborative work, and the dedication and shared vision of contributors who continue to inspire this evolving field. Such partnerships cultivate meaningful dialogue, expand global networks, and strengthen a dynamic platform for research exchange and innovation in resilient design.

About the Book

An innovative and multidisciplinary investigation into how the built environment might adjust to and lessen the effects of climate change is presented in Design Innovation for Climate-Resilient Built Environments. By using an integrated lens, i.e., exploring multi-scalar innovations from the city to the street, the building to the block, and the community to the region, this book bridges the fields of architecture, urban design, landscape architecture, urban planning, real estate, and construction management. Hence, it challenges traditional silos that exist in our built environments.

Cities and buildings have always been shaped by climate. Long before the phrase *climate resilience* appeared in policy documents, people were already experimenting with shade, airflow, materials, and landscape to survive heat, wind, drought, and storms. The first part of this book steps into that long lineage of knowledge and asks a simple but demanding question: how can design intelligence guide architecture and urban form in an era of accelerating climate uncertainty?

Part I, **"Design Intelligence for Climate-Resilient Architecture and Urban Form,"** establishes the conceptual ground for this exploration. Rather than approaching resilience only through engineering fixes or technological retrofits, this section frames resilience as a design mindset, one that draws from vernacular wisdom, ecological principles, and contemporary architectural innovation. The focus is on how the built environment can become adaptive, responsive, and regenerative, capable of evolving alongside changing climatic conditions.

If Part I deals with the *intelligence of design*, Part II turns to the machinery that allows that intelligence to operate at scale. Climate resilience does not emerge from architectural thinking alone. It depends on systems, i.e., technological, infrastructural, institutional, and urban, that shape how buildings are constructed, how cities function, and how societies adapt to environmental change.

Part II, **"Technological, Infrastructural, and Systemic Pathways to Climate Resilience,"** explores how innovation across multiple sectors is redefining the way resilience is embedded within the built environment. The focus shifts from spatial principles toward the operational frameworks that enable climate-responsive environments to function effectively over time. In other words, the conversation moves from *how we design* to *how systems perform, interact, and evolve* within complex urban ecosystems.

More importantly, this book sets itself apart as a roadmap for practitioners, academics, and decision-makers looking for impudent and feasible ideas for the future of climate-resilient cities and buildings by combining innovative design thinking with technology understanding and community-centered action. It serves as both a manual for defining resilience and a guide for creatively designing it.

Contents

Designing for a Planet in Flux: Architecture, Technology, and the Future of Climate-Resilient Cities 1
Ali Cheshmehzangi, Jian Zuo, Ayyoob Sharifi, Rongpeng Zhang, Abbas Ziafati Bafarasat, and Jie Zhao

Design Intelligence for Climate-Resilient Architecture and Urban Form

Passive and Nature-Based Design Innovations for Climate Resilient Cities: A Systematic Literature Review 11
Hadi RezaeiRad and Saba Sheikhi

The Key to Yazd's Resilience: Badgirs and Qanats, Innovative Secrets for Coping with the Arid Climate 57
Saeedeh Moayedfar and Seyedeh Saeedeh Hoseinyzadeh

Geometry-Led Residential Design for Climate Resilience in Hot Semi-arid Cities: A Case Study of Lahore 75
Ayesha Mehmood Malik, Saima Gulzar, and Memoona Rashid

Aqua Vascular Tower: A Biomimetic Approach to Climate-Resilient Vertical Architecture in Dense Urban Settings 95
Sadia-E-Jannat, Md Imamul Alam Ishat, and Nadia Monzur

Transforming Industrial Heritage into Climate-Resilient Cultural Spaces: Comparative Insights from Coal Drops Yard (UK) and Argo Factory (Iran) 121
Laleh Ramezani and Mahdiar Nezam

Housing Futures: Comparative Strategies for Climate-Resilient Housing in the Global North and South 141
Parsa Ahmadi Dehrashid and Hossein Mansourian

Technological, Infrastructural, and Systemic Pathways to Climate Resilience

AI-Enabled Emerging Technologies for High-Performance Building Benchmarking: Balance Energy, Comfort, and Health Resilience 177
Tian Li, Yunyang Ye, and Yi Lu

Integrating Source-Sink Theory and Road Pavement Solar Collectors for Urban Climate Resilience 193
Hadis Allahyari

Potential Strategies for Green Building Technology Adoption: Empirical Insights for Climate-Resilient Urban Development in Construction Sector 211
Bilal Hussain, Rui Feng Zhao, Syed Asif Ali Naqvi, and Atiq ur Rehman

Furthering a Systemic Approach to Resilience: Planning, Infrastructure, and Applications of Adaptation 225
Ling Almoubayyed, Alec Barnes, Megan Campbell, Jenny Hamel, Bethanie Martin, Alicia Adams, and Katrina Kelly-Pitou

Designing Resilience in Practice: Lessons, Transitions, and the Future of Climate-Responsive Cities 251
Ali Cheshmehzangi, Jian Zuo, Ayyoob Sharifi, Rongpeng Zhang, Abbas Ziafati Bafarasat, and Jie Zhao

Index 257

Editors and Contributors

About the Editors

Ali Cheshmehzangi is a Professor and Head of the School of Architecture, Design and Planning (ADP) at The University of Queensland. He has been in the World's top 2% field leader since 2021, recognized by Stanford University. He is among the top 20 global scholars in the urban sustainability research area. With a career spanning over two decades, he has made significant contributions to the academic and professional communities, with a focus on sustainable and environmentally conscious design. So far, Ali has published over 650 journal papers, articles, conference papers, book chapters, and reports. He also has 60 other academic books, some of which have received awards at the national, provincial, and municipal levels. He also received international awards and recognition for his research on urban resilience studies and sustainability research, as well as a Vice-Chancellor's award for his impactful contribution to higher education.

Jian Zuo is a Professor in Sustainable Construction at the School of Civil Engineering and Construction Management at Adelaide University. His main research interests are related to the sustainable built environment via both technical and managerial means. In particular, he has undertaken extensive research in construction and demolition waste recycling and management. As the lead researcher, he has been involved in a series of competitive grants in Australia such as Discovery Projects and Linkage Projects funded by the Australian Research Council as well as many other national key projects. He has long been engaged in the study of ecological towns, green buildings, and sustainable development of the construction industry. Published more than 100 papers in top journals, including more than 80 journal papers indexed by SCI and SSCI.

Ayyoob Sharifi is a Professor at the IDEC Institute, Hiroshima University. He also is a core member of the Network for Education and Research on Peace and Sustainability (NERPS). His research is mainly focused on urban sustainability, resilience, and climate change mitigation and adaptation. He has published extensively on these topics and is ranked among the world's top 1000 influential authors in all fields according to Stanford/Elsevier's Top 2% Scientist Rankings in 2024. Additionally, he received Clarivate's Highly Cited Researcher Award in 2024. Ayyoob actively contributes to global change research programs such as the Future Earth and has served as a lead author for the Sixth Assessment Report (AR6) of the Intergovernmental Panel on Climate Change (IPCC). The goal of his education and research activities is to inform actions toward building sustainable and peaceful communities.

Rongpeng Zhang is a Full Professor and dedicated PhD Supervisor at the School of Architecture and Planning, Hunan University. His primary research pursuits encompass building performance modeling and simulation for intelligent buildings, alongside advanced built environment control for healthy buildings. He was a primary developer of several pivotal modules within the DOE-endorsed EnergyPlus building simulation engine, including VRF-HR, fault detection, and demand response modules. He was a primary developer of Commercial Building Energy Saver (CBES) platform, which obtained the 2019 R&D 100 Award as one of the 100 most significant new technologies in the world. Before joining Hunan University, he led the building science team and conducted a series of research on human-centric built environment control at Well Living Lab, which obtained the 2018 Edison Award. He also served as an assistant professor at Mayo Clinic, and a postdoctoral fellow in the Building Technology and Urban Systems Division at Lawrence Berkeley National Laboratory. He received his Doctorate in Building Performance and Diagnostics from School of Architecture, Carnegie Mellon University and his B.S. and M.S. degrees in Built Environment and Facilities Engineering at Tongji University.

Jie Zhao is a lecturer of architecture in the Weitzman School of Design at the University of Pennsylvania, and President of P3Global, a Dubai-based sustainability, health and smart building advisory firm. He served as Head of Delos Labs and an Executive Vice President at Delos. His research focuses on Human-building Interactions, including healthy buildings, total building performance, occupant behavior in buildings, sensing and controls in buildings, health, comfort, and productivity of building occupants, as well as building energy efficiency and sustainability. He has published 20+ academic articles and serves as reviewers for 10+ top tier academic journals and conferences in the building science field. He frequently contributes to public media as a subject matter expert in healthy buildings, such as Wall Street Journal, CBS News, Forbes, and Money. Jie received his PhD in Building Performance and Diagnostics from Carnegie Mellon University and Master's and Bachelor's degrees from Tongji University.

All six editors have previously co-edited and published two books together on "*Designing Healthy Cities*" and "*Designing Healthy Buildings and Communities.*"

Abbas Ziafati Bafarasat is a Charted Town Planner (MRTPI), Fellow of the Higher Education Academy (FHEA), and Senior Lecturer in Town Planning, School of the Built Environment, Oxford Brookes University, where he also leads the Certificate in Spatial Planning Studies. Abbas undertook his PhD in Planning and Landscape at the University of Manchester, and his Post Doc at Technical University of Dortmund. His research interests involve healthy places, strategic planning, and education for sustainable development, with over 40 of his research outputs published in leading scientific platforms, mostly setting new directions (e.g., SIY Urbanism, Strategic Urban Design) and exploring novel horizons (e.g., bottom-up agenda setting) for addressing complex socio-environmental problems.

Contributors

Alicia Adams IMPACT (Integrated Multidisciplinary Performance Analytics and Climate-impact Team), SmithGroup, USA

Parsa Ahmadi Dehrashid Department of Human Geography and Planning, University of Tehran, Tehran, Iran

Hadis Allahyari Department of Physics and Astronomy "Augusto Righi," University of Bologna, Bologna, Italy

Ling Almoubayyed IMPACT (Integrated Multidisciplinary Performance Analytics and Climate-impact Team), SmithGroup, USA

Abbas Ziafati Bafarasat School of the Built Environment, Oxford Brookes University, Oxford, UK

Alec Barnes IMPACT (Integrated Multidisciplinary Performance Analytics and Climate-impact Team), SmithGroup, USA

Megan Campbell IMPACT (Integrated Multidisciplinary Performance Analytics and Climate-impact Team), SmithGroup, USA

Ali Cheshmehzangi School of Architecture, Design and Planning (ADP), The University of Queensland, Brisbane, Australia;
School of Architecture and Planning, Hunan University, Changsha, China

Saima Gulzar School of Architecture and Planning, University of Management and Technology, Lahore, Pakistan

Jenny Hamel IMPACT (Integrated Multidisciplinary Performance Analytics and Climate-impact Team), SmithGroup, USA

Seyedeh Saeedeh Hoseinyzadeh Department of Art and Architecture, Payame Noor University, Tehran, Iran

Bilal Hussain Faculty of Management and Economics, Kunming University of Science and Technology, Kunming, 650000 China;
Department of Economics, Government College University Faisalabad, Faisalabad, 38000 Pakistan

Md Imamul Alam Ishat Department of Mechanical Engineering, Chittagong University of Engineering & Technology, Chittagong, Bangladesh

Katrina Kelly-Pitou IMPACT (Integrated Multidisciplinary Performance Analytics and Climate-impact Team), SmithGroup, USA

Tian Li College of Architecture, University of Nebraska—Lincoln, Lincoln, NE, USA

Yi Lu Tepper School of Business, Carnegie Mellon University, Pittsburgh, PA, USA

Ayesha Mehmood Malik School of Architecture, Faculty of Arts and Architecture, University of Lahore, Lahore, Pakistan

Hossein Mansourian Department of Human Geography and Planning, University of Tehran, Tehran, Iran

Bethanie Martin IMPACT (Integrated Multidisciplinary Performance Analytics and Climate-impact Team), SmithGroup, USA

Saeedeh Moayedfar Department of Geography, Faculty of Humanities, Meybod University, Meybod, Iran

Nadia Monzur Department of Architecture, Rajshahi University of Engineering & Technology, Rajshahi, Bangladesh

Syed Asif Ali Naqvi Department of Economics, Government College University Faisalabad, Faisalabad, Pakistan

Mahdiar Nezam Manchester School of Architecture, Manchester, UK

Laleh Ramezani School of the Built Environment, Oxford Brookes University, Oxford, UK

Memoona Rashid School of Architecture, Faculty of Arts and Architecture, University of Lahore, Lahore, Pakistan

Atiq ur Rehman Department of Economics, Government College University Faisalabad, Faisalabad, Pakistan

Hadi RezaeiRad Department of Urbanism, Bu-Ali Sina University, Hamedan, Iran

Sadia-E-Jannat Department of Architecture, Rajshahi University of Engineering & Technology, Rajshahi, Bangladesh

Ayyoob Sharifi School of Architecture, Design and Planning (ADP), The University of Queensland, Brisbane, Australia;
The IDEC Institute and Network for Education and Research On Peace and Sustainability (NERPS), Hiroshima University, Hiroshima, Japan

Saba Sheikhi Department of Urban and Regional Planning, Tarbiat Modares University, Tehran, Iran

Yunyang Ye Department of Civil and Environmental Engineering, Colorado School of Mines, Golden, CO, USA

Rongpeng Zhang School of Architecture and Planning, Hunan University, Changsha, China

Jie Zhao The Weitzman School of Design, The University of Pennsylvania, Philadelphia, USA;
P3Global, Dubai, UAE

Rui Feng Zhao Faculty of Management and Economics, Kunming University of Science and Technology, Kunming, China

Jian Zuo School of Civil Engineering and Construction Management, Adelaide University, Adelaide, Australia

Designing for a Planet in Flux: Architecture, Technology, and the Future of Climate-Resilient Cities

Ali Cheshmehzangi, Jian Zuo, Ayyoob Sharifi, Rongpeng Zhang, Abbas Ziafati Bafarasat, and Jie Zhao

1 Climate Change, Urbanization, and the Imperative for Design Innovation

Cities are increasingly at the centre of the climate crisis. In addition to consuming enormous quantities of resources and having a major negative impact on the environment, they concentrate populations, infrastructure, and economic activity [1–3]. Rising temperatures, more intense heatwaves, urban flooding, water scarcity, and extreme weather events are no longer far-off threats but rather current realities influencing how cities operate and how people live in urban areas. Urban infrastructure and buildings, which were frequently built for historical climates, are now subject to stresses they were never meant to endure.

A. Cheshmehzangi (✉) · A. Sharifi
School of Architecture, Design and Planning (ADP), The University of Queensland, Brisbane, Australia
e-mail: A.Chesh@uq.edu.au

J. Zuo
School of Civil Engineering and Construction Management, Adelaide University, Adelaide, Australia

A. Sharifi
The IDEC Institute and Network for Education and Research On Peace and Sustainability (NERPS), Hiroshima University, Hiroshima, Japan

A. Cheshmehzangi · R. Zhang
School of Architecture and Planning, Hunan University, Changsha, China

A. Ziafati Bafarasat
School of the Built Environment, Oxford Brookes University, Oxford, UK

J. Zhao
The Weitzman School of Design, The University of Pennsylvania, Philadelphia, USA

P3Global, Dubai, UAE

A. Cheshmehzangi et al. (eds.), *Design Innovation for Climate-Resilient Cities and Built Environments*, Urban Sustainability, https://doi.org/10.1007/978-981-92-2297-1_1

As a result, the built environment is situated at a crucial junction of climate responsiveness and vulnerability. Cities' ability to withstand shocks, adjust to shifting environmental conditions, and lessen their ecological imprint is shaped by a combination of architecture, urban design, planning systems, and building technology [4, 5]. Climate resilience is now a design, technology, and governance problem rather than just an environmental goal.

Conventional methods of urban development have frequently divided technology, infrastructure, and design into separate fields. Planners concentrate on land use, engineers on structural performance, architects on spatial shape and aesthetics, and technology companies on operational effectiveness. However, a more comprehensive approach is required due to the intricacy of climate change. Isolated actions are insufficient to create resilient cities. They necessitate coordinated innovations that function at several scales, ranging from buildings and materials to neighbourhoods, infrastructures, and regional systems.

This book responds to this challenge by bringing together interdisciplinary perspectives that explore how the built environment can evolve under climate uncertainty. The book looks at how emerging technologies, infrastructure, architecture, and urban planning may all work together to create urban settings that are climate-responsive, low-carbon, and adaptable.

The book's central thesis is that climate resilience involves more than just fending off climatic stress; it also entails changing the way cities are planned, built, and run. Urban areas and buildings must be able to adjust to changing climate conditions while promoting social well-being, environmental sustainability, and human health. This calls for both innovative design principles and reliable technology systems that allow for long-term, durable performance.

This change is examined from complementary angles in the chapters that make up this volume. Some concentrate on the architectural and spatial intelligence that is incorporated into building and urban environment design. Others look at planning frameworks, infrastructure networks, and technology systems that allow resilience to function at scale. This collection of works shows how innovative design, technological development, and systemic urban thinking come together to provide climate resilience.

As a result, the book is divided into two complimentary sections. In the first, climate-responsive architecture and urban shape are based on design intelligence. The second looks at the systemic, technological, and infrastructure paths that enable the implementation and scaling of these design concepts in actual urban settings. The book provides both theoretical understanding and useful strategies for developing climate-resilient cities within this framework.

2 Design Intelligence for Climate-Resilient Architecture and Urban Form

Part I, **"Design Intelligence for Climate-Resilient Architecture and Urban Form,"** establishes the conceptual ground for this exploration. Rather than approaching resilience only through engineering fixes or technological retrofits, this section frames resilience as a design mindset, one that draws from vernacular wisdom, ecological principles, and contemporary architectural innovation. The focus is on how the built environment can become adaptive, responsive, and regenerative, capable of evolving alongside changing climatic conditions.

The chapters in this part unpack the design logics that allow buildings and urban spaces to interact more intelligently with their environmental context. Traditional climatic knowledge offers a valuable starting point. Across cultures and geographies, vernacular architecture developed sophisticated strategies to moderate heat, capture breezes, manage water, and create microclimates that support human comfort. These approaches remind us that climate-responsive design is not a new invention, but a rediscovery and reinterpretation of principles embedded in centuries of spatial practice.

But today's problem is more intricate. Urban environments that frequently act against natural processes rather than with them are the result of rapid urbanization, material development, and changing weather patterns. Therefore, integrated tactics that incorporate ecological thinking, technical innovation, and spatial creativity are required for modern architectural and urban design [6]. This is the point at which the concept of design intelligence becomes essential. The ability of architecture and urban form to anticipate environmental constraints, adapt dynamically to climatic circumstances, and create environments that promote long-term resilience is known as design intelligence.

Principles rooted in nature, such as nature-based principles or solutions, are essential to this change [7]. Porous surfaces, climate-responsive building envelopes, water-sensitive landscapes, and urban vegetation are examples of how design may collaborate with ecological processes instead of imposing inflexible structures on them. These methods show how architectural thinking has changed from manipulating nature to working with it. Within this concept, resilience appears not as a defensive tactic but as a generative one, whereby structures and cities change through adaptive interactions with their environment.

The potential of climate-responsive design is simultaneously increased by modern innovation. Designers may now test, improve, and optimize spatial solutions in previously unthinkable ways because of developments in digital modelling, environmental simulation, and material technology. Architects and planners may create environments that adapt to heat stress, flooding threats, wind dynamics, and energy demands with the aid of these technologies, which assist in converting climate data into practical design decisions.

Ultimately, this part of the book makes the case that design logic itself needs to incorporate climate resilience. Urban areas and buildings should be viewed as dynamic systems that can learn, adapt, and regenerate rather than just withstand the effects of climate change. The chapters in this section lay the conceptual groundwork for a new generation of climate-responsive design that redefines the relationship between cities, architecture, and the changing planet by fusing traditional knowledge, ecological thinking, and architectural innovation.

The chapters examine various design paths under this paradigm. Innovations in passive and nature-based architecture that help cities better adapt to environmental stress are reviewed in Chapter 2. In Chapter 3, the medieval urban fabric of Yazd is revisited, and it is examined how traditional elements like badgirs and qanats provide timeless lessons for dealing with arid weather. In order to better understand how spatial layout can improve climatic performance, Chapter 4 examines geometry-led residential design techniques for hot semi-arid environments. In Chapter 5, new strategies for climate-responsive high-rise design are presented, along with an introduction to biomimetic thinking in vertical architecture. While Chapter 7 looks at housing solutions in various global contexts and considers how residential architecture might solve socioeconomic and climatic concerns, Chapter 6 discusses the adaptive reuse of industrial legacy as climate-resilient cultural infrastructure. Together, these chapters illustrate how design intelligence can shape architecture and urban form in ways that align with environmental processes and support resilient urban futures.

3 Technological, Infrastructural, and Systemic Pathways to Climate Resilience

Resilience cannot be attained just through spatial thinking, even though design offers the conceptual basis for climate-responsive environments. Cities and buildings function through intricate networks of infrastructures, governance structures, and technology that affect their long-term performance. Because of this, the operational and systemic aspects of climate resilience are the focus of the book's second section.

Part II, **"Technological, Infrastructural, and Systemic Pathways to Climate Resilience,"** explores how innovation across multiple sectors is redefining the way resilience is embedded within the built environment. The focus shifts from spatial principles towards the operational frameworks that enable climate-responsive environments to function effectively over time. In other words, the conversation moves from how we design to how systems perform, interact, and evolve within complex urban ecosystems.

The way cities and buildings react to environmental stresses is changing due to technological advancements. Urban settings may now function with a level of awareness and adaptability that was previously unthinkable because of developments in smart building systems, climate-responsive materials, environmental monitoring, and digital modelling. Buildings can communicate with larger energy and infrastructure networks, control energy use, and react dynamically to variations in temperature and humidity thanks to these technologies. Simultaneously, digital tools like artificial intelligence, urban modelling environments, and simulation platforms are changing how planners and designers anticipate climate hazards and test adaptive tactics before putting them into practice in the real world.

Resilience at the urban level increasingly depends on the integration of cross-sectoral systems and infrastructures. Urban landscapes, transportation infrastructures, water systems, and energy networks all need to work together rather than separately. Heatwaves, flooding, resource scarcity, and infrastructural stress are examples of climate concerns that rarely affect a single aspect of the city. Across linked systems, they cascade. Therefore, coordinated strategies that connect technology innovation with planning frameworks that can direct long-term urban development are necessary for climate resilience.

Thus, the systemic aspect of resilience is covered in this section. Adaptive cities cannot be ensured by technological innovation alone unless it is integrated into sector-wide adoption paths, planning processes, and governance frameworks. The speed at which new technologies are incorporated into routine operations and the degree to which they support urban resilience objectives depend heavily on policies, standards, and institutional frameworks. As a result, the chapters in this section of the book consider the larger operational context in which technological innovation takes place.

Important societal issues are also brought up by technical advancement. Opportunities and conflicts are unavoidably created by the shift to low-carbon systems and climate-responsive infrastructures. Although emerging technologies have the potential to lower emissions, enhance environmental performance, and increase adaptability, they may also result in new types of infrastructure dependency, unequal access, or economic instability. As a result, the concept of a fair transition takes centre stage in the conversation. The pursuit of climate resilience must consider social justice, economic inclusiveness, and the wider societal effects of technology advancement.

Ultimately, Part II presents climate resilience as a systemic endeavour. For cities to be able to anticipate, absorb, and adapt to environmental stress, buildings, infrastructures, digital systems, and governance structures must collaborate. The built environment can transition from reactive responses to climate impacts to proactive systems that help cities navigate an increasingly uncertain future through technology innovation, integrated infrastructure, and forward-looking planning frameworks.

4 Towards Integrated Pathways for Climate-Resilient Cities

When combined, the two sections of this book offer complementary viewpoints on how the built environment might adapt to climate change. The conceptual and spatial techniques required to create climate-responsive architecture and urban form are provided by design intelligence. The operational capability needed to apply and maintain these methods in complicated urban environments is provided by technological and infrastructure systems.

This volume's contributions show that climate resilience is neither solely architectural nor solely technological. Design thinking, environmental knowledge, technology innovation, and institutional frameworks interact to produce it. Therefore, resilient cities must be viewed as dynamic systems where communities, infrastructure, buildings, and landscapes constantly interact.

The chapters also highlight the significance of multi-scalar thinking. At the scales of materials, structures, communities, and regional systems, climate resilience functions concurrently. If solutions created at one scale are to provide significant results, they must be in line with larger metropolitan processes. For instance, a building intended for passive cooling needs to be connected to water-sensitive landscapes, urban vegetation networks, and robust infrastructure systems that provide city-scale climate adaptability.

The understanding that social and cultural dimensions must continue to be taken into consideration for climate resilience is equally crucial. It is impossible to apply technological solutions consistently across cities or regions without taking governance systems, cultural norms, and economic situations into account. Therefore, inclusive planning procedures, community involvement, and local expertise are crucial elements of resilient urban development.

The aim of this book is to contribute to a broader conversation about how architecture, urban design, and urban technologies can collectively shape more adaptive and sustainable cities. It aims to stimulate fresh ideas on the connection between design innovation, technology advancement, and climate resilience by bringing together a variety of viewpoints and methods.

As climate uncertainty continues to intensify, the role of the built environment will become increasingly critical [8, 9]. The decisions taken now regarding infrastructure development, urban planning, and design will influence how cities react to environmental stresses in the next decades. Cities can be reimagined as platforms for innovation, adaptation, and regenerative urban futures in addition to being locations of vulnerability through interdisciplinary collaboration and integrated thinking.

References

1. Bigio AG (2003) Cities and climate change. In: Building safer cities: The future of disaster risk, pp 91–100.
2. Rees WE (2012) Cities as dissipative structures: global change and the vulnerability of urban civilization. In: Sustainability science: The emerging paradigm and the urban environment. Springer, New York, pp 247–273
3. Baker JL (ed) (2012) Climate change, disaster risk, and the urban poor: cities building resilience for a changing world. World Bank Publications, Washington D.C.
4. Roaf S, Crichton D, Nicol F (2009) Adapting buildings and cities for climate change: a 21st century survival guide. Routledge, New York and London

5. Morel H, Dorpalen BD (2023) Adaptive thinking in cities: urban continuity within built environments. Climate 11(3):54
6. Yunxuan W, Ruikai Y, Ibrahim NLBN (2025) From traditional to modern: Cultural integration and innovation in sustainable architectural design education. Journal of Ecohumanism 4(1):2079–2093
7. Cohen-Shacham E, Andrade A, Dalton J, Dudley N, Jones M, Kumar C, Maginnis S, Maynard S, Nelson CR, Renaud FG, Welling R (2019) Core principles for successfully implementing and upscaling Nature-based Solutions. Environ Sci Policy 98:20–29
8. Wilby RL (2007) A review of climate change impacts on the built environment. Built environment 33(1):31–45
9. Zięba Z, Dąbrowska J, Marschalko M, Pinto J, Mrówczyńska M, Leśniak A, Kazak JK (2020) Built environment challenges due to climate change. In: IOP conference series: earth and environmental science, vol 609, No 1. IOP Publishing, p 012061

[illegible] (2009) [illegible]

Design Intelligence for Climate-Resilient Architecture and Urban Form

Passive and Nature-Based Design Innovations for Climate Resilient Cities: A Systematic Literature Review

Hadi RezaeiRad and Saba Sheikhi

1 Introduction

The rapid pace of urbanization, coupled with the intensifying impacts of climate change, has positioned cities at the epicenter of global environmental challenges. Urban environments experience amplified climate risks through phenomena such as urban heat islands, increased frequency of extreme weather events, elevated air pollution levels, and compromised water management systems. These challenges necessitate a fundamental transformation in how cities are planned, designed, and managed to ensure long-term sustainability and resilience.

The concept of climate resilience in urban contexts encompasses the capacity of urban systems to anticipate, prepare for, respond to, and recover from climate-related disruptions while maintaining essential functions and adapting to changing conditions. Traditional approaches to urban climate adaptation have often relied on engineered solutions and energy-intensive technologies. However, growing recognition of the limitations and environmental costs of such approaches has catalyzed interest in alternative strategies that work with natural processes and passive design principles.

Passive design strategies represent a paradigm shift toward minimizing energy consumption and environmental impact through thoughtful architectural and urban design decisions. These approaches harness natural forces such as solar radiation, wind patterns, and thermal dynamics to create comfortable built environments without relying heavily on mechanical systems. Key passive design elements include strategic building orientation, thermal mass utilization, natural ventilation systems, daylight optimization, and responsive building envelopes. The effectiveness

H. RezaeiRad (✉)
Department of Urbanism, Bu-Ali Sina University, Hamedan, Iran
e-mail: H.rezaeirad@Basu.ac.ir

S. Sheikhi
Department of Urban and Regional Planning, Tarbiat Modares University, Tehran, Iran

A. Cheshmehzangi et al. (eds.), *Design Innovation for Climate-Resilient Cities and Built Environments*, Urban Sustainability, https://doi.org/10.1007/978-981-92-2297-1_2

of passive design in reducing energy consumption, improving indoor environmental quality, and enhancing occupant comfort has been well-documented across various climatic contexts.

Complementing passive design approaches, nature-based solutions have emerged as a comprehensive framework for addressing urban climate challenges through the strategic use of natural and semi-natural features. The International Union for Conservation of Nature defines NBS as actions to protect, sustainably manage, and restore natural or modified ecosystems that address societal challenges effectively and adaptively, simultaneously providing human well-being and biodiversity benefits. In urban contexts, NBS encompass a diverse range of interventions including green infrastructure, urban forests, constructed wetlands, permeable surfaces, and integrated water management systems.

The significance of integrating passive design and nature-based solutions extends beyond their individual contributions to climate resilience. These approaches offer multiple co-benefits including biodiversity conservation, air quality improvement, social cohesion enhancement, economic development, and health promotion. They align with broader sustainability objectives and contribute to achieving multiple sustainable development goals simultaneously.

Despite the growing body of literature on passive design and nature-based solutions, several research gaps persist. Most existing studies remain fragmented, discipline-specific, or focused on individual strategies, with a notable lack of comprehensive and systematic syntheses that critically integrate empirical evidence across passive design and nature-based solutions. First, there remains limited comprehensive synthesis of the effectiveness of different passive and NBS strategies across diverse urban contexts and climatic conditions. Second, the integration challenges and barriers to implementation are often context-specific and require systematic analysis to identify common patterns and solutions. Third, the long-term performance and maintenance requirements of these interventions need better documentation and understanding. Finally, the synergistic effects of combining passive design with nature-based solutions remain underexplored, limiting the development of holistic resilience strategies.

The purpose of this systematic literature review is to critically examine and synthesize the current state of knowledge regarding passive and nature-based design innovations for climate-resilient cities. Specifically, this review aims to: (1) identify and categorize the range of passive design strategies and nature-based solutions implemented in urban environments; (2) evaluate the empirical evidence for their effectiveness in enhancing climate resilience; (3) analyze the socio-environmental co-benefits and trade-offs associated with these interventions; (4) examine the institutional, regulatory, and practical challenges hindering their widespread implementation; and (5) identify research gaps and future directions for advancing climate-resilient urban design.

The chapter is structured as follows. Section 2 outlines the systematic review methodology, including research questions, search strategy, selection criteria, and analytical framework. Section 3 presents the results of the literature analysis, providing a descriptive overview of selected studies and synthesizing findings related

Table 1. Research questions guiding the systematic literature review

RQ	Research Question
RQ1	What are the main categories of passive design strategies and nature-based solutions being implemented in urban environments to enhance climate resilience?
RQ2	What empirical evidence exists regarding the effectiveness of these strategies in addressing specific climate challenges such as urban heat islands, stormwater management, and energy consumption?
RQ3	What are the documented socio-environmental co-benefits and potential trade-offs associated with passive design and nature-based interventions in urban contexts?
RQ4	What institutional, regulatory, and practical barriers hinder the widespread implementation of passive and nature-based design innovations in cities?
RQ5	How do passive design strategies and nature-based solutions interact synergistically, and what frameworks exist for their integrated implementation?

to each research question. Section 4 discusses the implications of the findings and exploring theoretical and practical applications. Section 5 acknowledges the limitations of the review, while Sect. 6 identifies future research directions. Finally, Sect. 7 provides a comprehensive conclusion summarizing the key contributions and recommendations for policy and practice.

2 Review Protocol/Methodology

2.1 Research Questions

This systematic literature review was guided by five primary research questions designed to examine passive and nature-based design innovations for climate-resilient cities (see Table 1):

2.2 Search Strategy

A comprehensive search strategy was developed to identify relevant literature published between January 2010 and July 2025. The search was conducted across three major databases: Scopus, Web of Science Core Collection, and ScienceDirect. These databases were selected for their comprehensive coverage of peer-reviewed literature in urban planning, environmental science, architecture, and related disciplines. The search strategy employed a combination of keywords organized into three main concept groups (see Table 2). Boolean operators (AND, OR) were used to combine search terms within and across concept groups. The search was limited to peer-reviewed journal articles published in English. Gray literature was excluded

Table 2. Concept groups and keywords used in the search strategy

Concept Group	Keywords
Urban and climate concepts	"Urban climate resilience," "climate adaptation," "climate change," "urban heat island," "extreme weather," "urban sustainability"
Design and solution concepts	"Passive design," "nature-based solutions," "green infrastructure," "natural ventilation," "thermal mass," "bioclimatic design," "sustainable architecture"
Implementation concepts	"Urban planning," "building design," "landscape architecture," "water management," "energy efficiency"

Table 3. Inclusion and exclusion criteria for literature selection

Criterion	Description
Inclusion	Peer-reviewed journal articles; Published between January 2010 and July 2025; Focus on urban environments or urban-scale applications; Empirical studies, case studies, or comprehensive reviews related to passive design or nature-based solutions; Clear connection to climate resilience, adaptation, or mitigation objectives; Sufficient methodological detail to assess study quality
Exclusion	Non-peer-reviewed publications (books, reports, dissertations, conference papers); Studies focused solely on rural or non-urban contexts; Purely theoretical papers without empirical evidence or practical applications; Studies primarily focused on traditional engineering solutions without passive or nature-based components; Publications not available in English; Duplicate publications or abstracts without full papers

to maintain consistency in quality standards, though this limitation is acknowledged in the review.

2.3 Inclusion and Exclusion Criteria

The inclusion and exclusion criteria applied in the literature selection process are summarized below (see Table 3). Inclusion criteria were established to ensure the relevance and quality of selected studies. Exclusion criteria eliminated studies that did not meet the review objectives.

2.4 Selection Process

The study selection process followed the PRISMA guidelines and consisted of multiple screening stages. Initial database searches yielded 8041 potentially relevant records. After removing duplicates (2292) in EndNote software, 5749 unique records remained for title and abstract screening. Two independent reviewers conducted

Table 4. Quality assessment criteria

Criterion	Description
Study design appropriateness	Alignment between research objectives and methodology
Data collection methods	Clarity and rigor of data collection procedures
Sample size and representativeness	Adequacy of sample size and generalizability of findings
Analytical approach	Appropriateness and transparency of analytical methods
Reporting quality	Completeness and clarity of reporting
Bias and limitations	Acknowledgment and assessment of potential biases and limitations

the title and abstract screening using the predetermined inclusion and exclusion criteria. Disagreements were resolved through discussion and consultation with a third reviewer when necessary. This process resulted in 508 articles selected for full-text review. Full-text screening was conducted by the same two reviewers, with detailed assessment against all inclusion and exclusion criteria. Articles were categorized as: (1) included for final analysis, (2) excluded with reasons, or (3) requiring further discussion. This process resulted in 224 studies meeting all criteria for inclusion in the final analysis.

2.5 *Quality Assessment*

A quality assessment framework was developed to evaluate the methodological rigor and reliability of included studies. The framework incorporated elements from established quality assessment tools adapted for the diversity of study types encountered in this review. Each study was assigned a quality score on a scale of 1–5 (low to high quality) based on these criteria. Studies scoring below 3 were excluded from the analysis, resulting in a final corpus of 161 high-quality studies for synthesis. Assessment criteria included (see Table 4):

2.6 *Data Extraction and Synthesis Method*

A standardized data extraction form was developed to systematically capture relevant information from each included study. Extracted data included (see Table 5):

he selection of studies for this systematic review followed the PRISMA 2020 guidelines, which involved multiple stages of identification, screening, eligibility assessment, and inclusion of final studies (see Fig. 1).

Data synthesis employed a mixed-methods approach combining quantitative analysis where possible with qualitative thematic analysis. Quantitative synthesis

Table 5. Data extraction framework

Data Item	Description
Bibliographic information	Author(s), publication year, journal, study location
Study characteristics	Research design, sample size, duration, context
Intervention details	Type of passive design strategy or nature-based solution, scale of implementation, design specifications
Outcomes measured	Climate resilience indicators, performance metrics, co-benefits assessed
Key findings	Primary results, effect sizes where available, statistical significance
Implementation factors	Barriers, facilitators, costs, maintenance requirements
Quality indicators	Methodological strengths and limitations

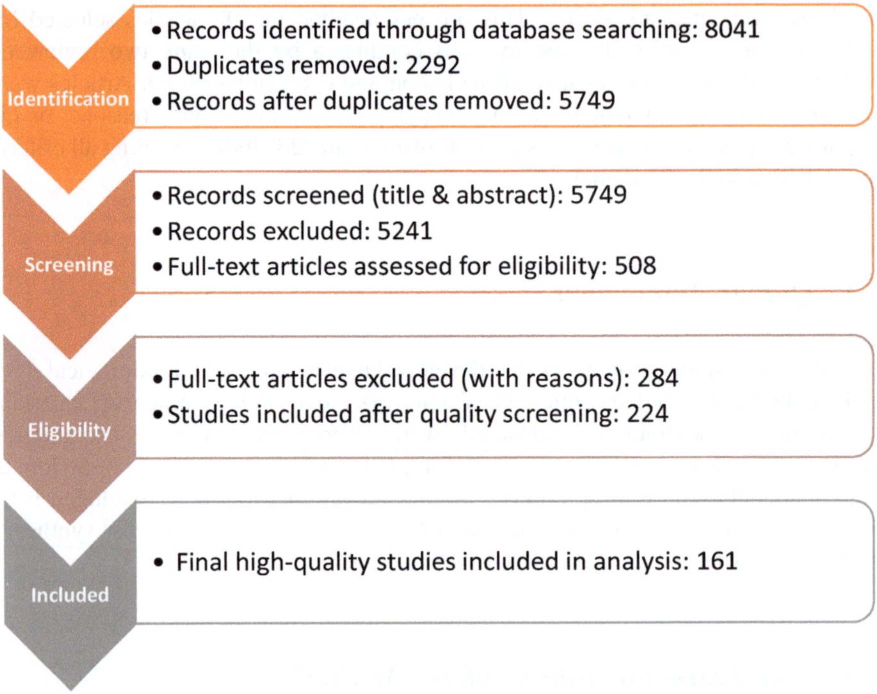

Fig. 1 PRISMA 2020 flow diagram of the study selection process

included meta-analysis of comparable outcome measures across studies with sufficient data quality and homogeneity. Qualitative synthesis utilized thematic analysis to identify patterns, themes, and relationships across studies, particularly for complex phenomena such as implementation barriers and co-benefits. The synthesis process was iterative, involving multiple rounds of coding and theme development. Initial coding was conducted by two reviewers independently, with subsequent discussion and refinement of themes. A final thematic framework was developed through consensus, organizing findings around the five research questions while allowing for emergent themes and cross-cutting issues.

3 Results

The results provide an overview of the reviewed studies and highlight key patterns across contexts, methods, and outcomes.

3.1 Descriptive Overview of Selected Studies

The final analysis included 161 peer-reviewed studies published between 2010 and 2025, representing a substantial and growing body of literature on passive and nature-based design innovations for climate-resilient cities. Publication trends showed a marked increase in research activity over the review period, with 62% of studies published after 2017, indicating heightened academic and practical interest in these approaches.

Geographically, the studies represented diverse urban contexts across six continents, with the highest concentration in Europe (34%), followed by North America (26%), Asia (22%), Australia (10%), South America (5%), and Africa (3%). This distribution reflects both research capacity and the urgency of climate adaptation needs in different regions. European studies were particularly prominent in nature-based solutions research, while North American studies showed strength in passive design strategies.

Methodologically, the included studies employed research approaches: experimental studies (28%), case study analyses (25%), modeling and simulation studies (23%), comparative analyses (15%), and mixed-methods approaches (9%). This diversity reflects the multidisciplinary nature of the field and the different scales at which passive and nature-based interventions operate.

The studies addressed multiple climatic challenges, with urban heat island mitigation being the most frequently studied (67% of studies), followed by stormwater management (45%), energy efficiency (43%), air quality improvement (31%), and carbon sequestration (18%). Many studies addressed multiple challenges simultaneously, reflecting the multi-functional nature of passive and nature-based solutions.

3.2 Categories of Passive Design Strategies and Nature-Based Solutions (RQ1)

The analysis revealed a rich taxonomy of passive design strategies and nature-based solutions being implemented in urban environments. These interventions were categorized into distinct but often overlapping groups based on their primary mechanisms and applications (see Tables 6 and 7). The synthesis indicates that strategies operating at multiple spatial scales and those integrated within broader urban systems tend to demonstrate more consistent and transferable performance across different climatic contexts. The values presented in the tables represent the average of all values reported in the reviewed articles.

3.3 Effectiveness in Addressing Climate Challenges (RQ2)

The empirical evidence regarding the effectiveness of passive design strategies and nature-based solutions in addressing specific climate challenges revealed both significant potential and important contextual dependencies (see Table 8). The synthesis indicates that the most pronounced and consistently reported impacts are associated with urban heat island mitigation and reductions in energy demand, where multiple interventions demonstrate convergent performance across diverse climatic contexts. In contrast, while outcomes related to stormwater management and disaster risk reduction are frequently identified as promising, the magnitude of these effects is supported by a smaller and more heterogeneous body of quantitative evidence.

3.4 Socio-environmental Co-benefits and Trade-Offs (RQ3)

The analysis revealed extensive socio-environmental co-benefits associated with passive design and nature-based solutions, though some studies also documented important trade-offs and unintended consequences (see Table 9). The reviewed evidence consistently highlights significant social and health benefits, including improvements in mental well-being, physical activity, thermal comfort, and equity outcomes, alongside substantial economic returns through energy savings, infrastructure cost reductions, and local economic development. At the same time, the analysis reveals that these benefits are accompanied by important trade-offs and unintended consequences such as increased maintenance demands, water consumption, social displacement risks, and land-use competition which can undermine long-term effectiveness if not proactively addressed.

Table 6. Summary of building- and urban-scale passive design strategies

Scale	Strategy	Key Findings	References
Building scale	Solar orientation & shading	Optimal orientation reduces cooling demand by 15–30% in hot climates; maximizes winter gains in temperate regions	[1, 89, 161, 198, 247]
	Thermal mass & phase change materials	High-mass materials moderate indoor temps; PCM integration stabilizes indoor temperature by 3–5 °C	[9, 89, 188, 213, 247]
	Natural ventilation	Cross/stack ventilation and courtyards cut cooling energy by 20–40%	[44, 45, 81, 112, 198, 247]
	Building envelope innovations	– Cool roofs reduce surface temps by 10–20 °C – Adaptive facades improve comfort and energy efficiency – Green Glass Space (GGS) acts as a buffer zone between outdoor and indoor environments, utilizes solar energy, improves thermal performance in both winter and summer, and reduces building energy use and CO_2 emissions	[24, 120, 124, 179, 200, 252]
	wind-induced evaporative cooling	Passive system combining wind flow with wetted permeable surfaces; reduced indoor temp by up to 3.9 °C, cooling capacity 9–476 W; effective in tropical hot–humid climates	[158, 198]
Urban scale	Urban morphology & density	Building height–street width ratios influence wind and heat dissipation	[94, 105, 134, 199, 215]
	Street orientation & canyon design	E–W orientation reduces solar gain in tropics; N–S orientation improves winter access in temperate climates	[3, 94, 144, 181, 198]
	Public space design	Shade, water features, and responsive materials lower ambient temps by 2–5 °C	[32, 130, 155, 197]

3.5 Implementation Barriers and Challenges (RQ4)

The analysis revealed systematic barriers hindering the widespread implementation of passive design and nature-based solutions across multiple categories and scales (see Table 10). The reviewed evidence highlights persistent regulatory misalignment,

Table 7. Summary of nature-based solutions

Category	Key Findings	References
Green Infrastructure	– Green roofs reduce peak summer temps by 1–4 °C (extensive) and provide 50–80% annual stormwater retention (intensive) – Green walls lower surface temps by 2–8 °C; impact depends on design & maintenance – Urban trees reduce surface temps by 2–15 °C; urban forests mitigate heat island effects – Permeable pavements reduce surface temps by 5–12 °C and improve infiltration	[15, 47, 61, 66, 69, 137, 145, 183, 186, 208, 217, 238, 239, Lui et al. 2025]

(continued)

Table 7. (continued)

Category	Key Findings	References
Water-Sensitive Urban Design (WSUD)	– Wetlands & bioswales improve water quality (70–95% pollutant removal) and provide cooling via evapotranspiration – Rainwater harvesting & blue-green systems enhance cooling and ecosystem services – Urban ponds/streams/fountains reduce temps by 1–6 °C within 50–200 m	[6, 7, 13, 34, 36, 41, 80, 103, 110, 174, 184, 185, 189, 206, 232, 248, 251, Lui et al. 2025]
Biodiversity and Ecosystem Services	– Pollinator gardens & native landscaping support biodiversity with low water needs – Monoculture plantations reduce biodiversity, but heterogeneous and structurally diverse plantations can enhance ecosystem functions and species richness – Urban agriculture & food forests provide food security, local cooling, and social benefits – Wildlife habitats & green corridors improve ecosystem resilience & climate regulation	[20, 37, 41, 49, 60, 61, 63, 83, 113, 128, 132, 135, 226, 227]

Table 8. Effectiveness of passive design strategies and nature-based solutions in addressing climate challenges

Domain	Strategy/Solution	Key Findings	References
Urban Heat Island Mitigation	Urban tree canopy & green infrastructure	Air temperature reduction 0.5–2.0 °C; surface temperature reduction 2–15 °C depending on canopy density & climate	[165, 189, 218, 254, Lui et al. 2025]
	Extensive green roofs	Cooling effects: 0.3–1.5 °C (building scale), 0.1–0.8 °C (neighborhood scale)	[8, 175, 200]
	Intensive green roofs	Enhanced performance but higher implementation & maintenance costs	[69, 160]
	Cool roofs	Surface temperature reduction 10–40 °C; ambient air temperature reduction 0.2–1.2 °C	[74, 133, 179, 194]
	Natural ventilation	Indoor temperature reduction 2–6 °C	[27, 78]
	Building orientation & shading	Moderate ambient temperature reduction	[11, 225]
Stormwater Management	Blue-Green roofs	Green roofs (extensive: 45–75%, intensive: 60–90%)—Retention; peak flow reduction 65–95% Blue-green variant: Retention ~ 67%; ~ 50% no runoff; full absorption in dry season; climate-sensitive performance	[15, 26, 69, 226, 230]
	Bioswales & constructed wetlands	Pollutant removal: Total Suspended Solids (70–95%), nutrients (60–90%), heavy metals (50–85%); peak flow reduction 40–80%	[8, 56, 109, 131, 162, 234, 243]
	Permeable pavements	Infiltration rates 2–10 × higher than conventional; performance decline without maintenance	[41, 114, 117, 142]
	Integrated blue-green systems	Combined systems outperform individual ones; promising for urban retrofits	[43, 172]

(continued)

Table 8. (continued)

Domain	Strategy/Solution	Key Findings	References
	Regional dependency	Effectiveness linked to rainfall, soil, infrastructure; major benefits in combined sewer systems	[69, 82, 176, 193]
Energy Efficiency & Carbon Reduction	Natural ventilation	Cooling energy savings 20–60%	[170, 204]
	Building envelope improvements (insulation, cool roofs, adaptive facades)	Energy savings 15–40%; higher impact in extreme climates	[5, 123, 196]
	Green roofs	Cooling load reduction 10–30% depending on coverage	[22, 42]
	Urban tree canopy	Neighborhood cooling load reduction 5–15%	[122, Lui et al. 2025]
	Solar orientation & building optimization	Energy demand reduction 20–40%; highly climate-dependent	[171]
	Thermal mass layout strategies	Layout optimization of walls and rooms improves thermal storage, yielding 0.5–3 °C indoor cooling and notable energy savings	[89]
	Integrated passive design strategies	Synergistic savings 50–80% vs. conventional	[93, 119, 224]
	Carbon reduction	Dependent on local energy grid; higher benefits in carbon-intensive grids; vegetation adds sequestration	[15, 166, 195]
Air Quality Improvement	Urban vegetation	Pollutant reduction (PM2.5, PM10, etc.) 10–40% with dense canopy	[2, 126, 205]
	Green walls & vertical gardens	Pollutant reductions 15–60% near surfaces; highly localized	[15, 69, 231, 240]
	Passive design (natural ventilation)	Indirect air quality gains via reduced energy & emissions	[38, 222, 250]

(continued)

Table 8. (continued)

Domain	Strategy/Solution	Key Findings	References
	Vegetation-pollution interactions	Potential negative impacts: reduced airflow, pollutant trapping in dense vegetation	[16, 33, 84, 201]
Disaster Risk Reduction	Nature-based approaches (e.g., upland reforestation, coastal wetlands, mangroves, integrated frameworks)	NBS reduce landslide and storm surge risks, enhance coastal resilience, and provide co-benefits (biodiversity, carbon storage, habitat). evidence shows ~ 70% cost-effectiveness, often outperforming gray infrastructure; aligns with Sendai Framework for integrated risk management	[8, 18, 41, 154, 185]

fragmented governance structures, and short-term financial and procurement frameworks as the most frequently cited constraints, often reinforcing one another across scales. In addition, limitations in professional knowledge, modeling tools, maintenance practices, and social acceptance further constrain scalability and long-term performance, contributing to a persistent gap between demonstrated effectiveness and real-world adoption.

4 Discussion and Implications

This section discusses the main findings of the review and their implications for theory and practice.

4.1 Interpretation of Findings

The systematic review reveals a mature and rapidly evolving field of research on passive design and nature-based solutions for climate-resilient cities. The evidence demonstrates substantial potential for these approaches to address multiple urban climate challenges simultaneously while providing significant co-benefits for human health, environmental quality, and economic development. However, the findings also reveal important implementation challenges that limit the translation of research knowledge into widespread practice.

Table 9. Socio-economic and environmental co-benefits and trade-offs of passive design strategies and nature-based solutions

Domain	Category/ Strategy	Key Findings	References
Social and health benefits	Mental health & well-being	-Access to quality green space improves stress reduction, mood, cognitive function; effects comparable to clinical interventions	[41, 52, 76, 97, 146]
	Physical health	-Increased physical activity (walking, cycling) 15–30% higher in areas with high-quality green infrastructure & walkable design -Reduced respiratory & cardiovascular illness linked to air quality gains; economic value \$50–500 per person/year -Green streets reduce traffic noise	[15, 34, 77, 97, 100, 104, 167, 242, 245]
	Community health & equity	Urban agriculture and green space access improve nutrition, reduce health disparities, and lower chronic disease mortality	[41, 97]
	Thermal comfort & heat risk	Cooling benefits protect vulnerable populations (elderly, no AC); reduced heat-related mortality & morbidity	[31, 95, 98, 106, 207]

(continued)

Table 9. (continued)

Domain	Category/ Strategy	Key Findings	References
	Aesthetic benefits	Green walls and facades enhance urban streetscapes and visual appeal; contribute to sense of place and perceived quality of living environments	[41, 97]
Economic benefits	Energy savings	-Passive design reduces energy costs with 3–15 year payback depending on type & local costs -Green roofs cut building cooling costs by 10–30%	[97, 102, 121, 185, 212]
	Green infrastructure Return on Investment (ROI)	-Longer payback (5–25 years) if only direct benefits; positive returns when co-benefits included -Government subsidies & tax credits improve ROI and financial attractiveness	[28, 97, 177]
	Property values	Property premiums of 3–20% for high-quality green/passive design areas; stronger in dense cities	[75, 77, 97, 118]

(continued)

Table 9. (continued)

Domain	Category/ Strategy	Key Findings	References
	Infrastructure cost savings	-Green stormwater systems 20–50% cheaper over lifecycle than gray infrastructure -Rain gardens & bioswales reduce stormwater runoff treatment costs by 30–50% -Green roofs and permeable surfaces extend building materials' lifespan, resulting in lower long-term maintenance costs -Rainwater harvesting & graywater recycling reduce municipal water demand and lower water costs -Green roofs & walls mitigate extreme temperature damage, lowering repair and renovation costs	[43, 57, 97, 107, 241, Lui et al. 2025, Şen et al. 2024]
	Local economic development	Job creation, tourism, business attractiveness; multiplier effects of 1.5–2.5	[65, 97, 153]

(continued)

Table 9. (continued)

Domain	Category/ Strategy	Key Findings	References
Environmental co-benefits	Biodiversity conservation	– Green infrastructure supports pollinators & birds – species diversity increased by 30–200% with native landscaping	[15, 19, 25, 88, 132, 221]
	Carbon sequestration	– Mature urban trees sequester 10–50 kg CO_2/tree/year – green roofs provide smaller but meaningful contributions	[15, 55, 92, 97, 127, 166, 219, 228]
	Water quality	40–80% reductions in nutrient & sediment loading; broader watershed benefits	[15, 136, 149, 180, 189]
	Microclimate regulation	Benefits beyond cooling: humidity control, wind modification, enhanced precipitation; especially important in arid climates	[41, 163, 211, 237]
Trade-offs and unintended consequences	Water consumption	Irrigation for green roofs/ landscaping may offset benefits in water-scarce regions; mitigated by drought-tolerant plants & efficient systems	[116, 156, 190, 249]

(continued)

Table 9. (continued)

Domain	Category/ Strategy	Key Findings	References
	Maintenance & costs	Nature-based solutions require 20–40% higher annual maintenance; performance degrades without protocols	[10, 71, 99, 169]
	Gentrification & equity	High-quality green amenities may trigger gentrification & displacement; need equitable policy frameworks	[17, 46, 140, 192, 244]
	Urban development vs. NBS	Rigid trade-offs between land for economic growth (buildings, parking, facilities) and land for NBS/ UGS; tensions between gray vs. green valuation systems; implications for equity and long-term resilience	[203, 216]
	Pests & allergens	Poor species selection can increase allergens or vector habitats	[39, 64, 139, 141]

(continued)

Table 9. (continued)

Domain	Category/ Strategy	Key Findings	References
	Functional trade-offs	Conflicts in design and use of NBS: setting priorities (stormwater vs. cooling, biodiversity vs. recreation), usage conflicts (CES vs. ecological functions), and recognition dilemmas (restoration vs. social benefits, regeneration vs. displacement)	[203, 216]
	Space competition	In dense cities, green infrastructure competes with housing, transport, commerce; requires integrated planning	[21, 86, 91, 191]
	Governance trade-offs	-Inclusive processes improve justice and participation but require high time/resources, delaying implementation -Making NBS more accessible increases social benefits but may reduce ecological value and create user conflicts	[41, 216]

Table 10. Implementation Barriers and Challenges (RQ4)

Category	Barrier / Challenge	Details	References
Institutional and Regulatory Barriers	Misalignment with codes & zoning	– Existing building codes and zoning regulations often conflict with sustainable design goals – Lack of provisions for green/passive strategies – Approval delays of 6–24 months	[54, 148, 151, 154, 202, 255]
	Permitting complexity	Integrated systems face challenges due to multiple agencies with fragmented expertise, increasing costs and timelines	[59, 187, 220]
	Liability & professional knowledge	– Architects/engineers hesitant due to liability and lack of local precedents – Professional licensing/education slow to update sustainable design knowledge	[140, 214]
	Limited collaborative governance	Development of NBS hindered by siloed governmental departments, fragmented responsibilities, different investment horizons among stakeholders, and dominance of engineering expertise over other knowledge domains	[48, 62, 140]
	Procurement limitations	Conventional cost-focused procurement overlooks long-term co-benefits of sustainable design	[51, 159, 168]

(continued)

Table 10. (continued)

Category	Barrier / Challenge	Details	References
Financial and Economic Barriers	High upfront costs	Green infrastructure shows 10–50% higher capital costs vs. conventional systems, especially difficult for affordable housing/ public projects	[97, 164, 182, 209]
	Financing misalignment	Conventional short-term financing fails to capture long-term sustainable design benefits	[97, 152, 154, 223]
	Split incentives	Investment vs. benefit misalignment, e.g., landlords invest while tenants benefit from energy savings	[23, 29, 125]
	Risk/insurance issues	Insurance markets fail to reward resilience benefits of sustainable designs	[111, 115, 138, 150]
	Low private sector engagement	Lack of profitable business models, long payback periods, short-term investment horizons, uncertainty about costs and performance, and low market demand reduce willingness of private sector actors to invest in NBS	[57, 62, 235]
	Valuation difficulties	Challenges in quantifying/ co-benefits; ecosystem service valuation inconsistent and underused	[14, 58, 87]
Technical and Knowledge Barriers	Expertise gaps	Limited training among professionals in integrated passive/ nature-based approaches; continuing education lagging	[12, 30, 35, 41, 73, 97, 113, 226, 235]

(continued)

Table 10. (continued)

Category	Barrier / Challenge	Details	References
	Modeling / terminology limitations	– Tools inadequate for predicting performance of integrated systems, reducing confidence and adoption – Ambiguity and inconsistency in terminology hinder adoption and integration into policy	[15, 108, 154, 165, 233, 253]
	Local adaptation gaps	Lack of local climate data/guidelines leads to suboptimal performance	[40, 79, 210, 227]
	Maintenance/ monitoring weaknesses	Poor protocols cause system degradation and failures, discouraging future adoption	[18, 72, 154, 241]
	Scalability & replication limitations	NBS effective at pilot scale, but face land, cost, and governance barriers when expanded city-wide	[97]
	Supply chain constraints	– Limited availability of native plants and specialized materials – Nursery/supplier capacity often inadequate	[173, 178, 230]
Social and Cultural Barriers	Public acceptance	Resistance due to pest, maintenance, or aesthetic concerns about visible green infrastructure	[4, 67, 68, 157, 229, 235]
	Cultural preferences	Preference for manicured landscapes and conventional designs hinders natural/integrated alternatives	[90, 129, 154, 236, 246]
	Risk aversion	Developers/clients prefer conventional, proven methods over innovative solutions	[53, 96, 101, 147]

(continued)

Table 10. (continued)

Category	Barrier / Challenge	Details	References
	Competition over urban space	Competition over limited urban space between NBS and other priorities (housing, commercial development, solar panels), influenced by both economic costs and social priorities	[62, 63, 203]
	Community engagement	Lack of resident participation undermines neighborhood-scale interventions; engagement underestimated in planning	[4, 43, 50, 70, 97, 143]

The effectiveness of passive design strategies and nature-based solutions shows strong context dependencies related to climate, urban morphology, and local socio-economic conditions. This suggests that successful implementation requires careful adaptation to local conditions rather than adoption of generic solutions. The most effective approaches appear to be those that integrate multiple strategies in comprehensive systems designed for specific local contexts.

The co-benefits of these approaches often exceed their direct climate adaptation benefits, suggesting that economic justification should incorporate multiple benefit streams rather than focusing solely on climate impacts. This finding has important implications for financing mechanisms and investment decision-making processes.

The implementation barriers identified in this review reflect systemic challenges in urban development processes that extend beyond technical design issues. Addressing these barriers requires coordinated efforts across multiple sectors and scales, from professional education and regulatory reform to community engagement and financing innovation.

4.2 Theoretical Implications

The findings contribute to several theoretical frameworks relevant to urban sustainability and climate adaptation. From a socio-technical systems perspective, the review highlights the importance of institutional and social factors in determining the success of technical interventions. Passive design and nature-based solutions cannot be understood purely as technical fixes but must be analyzed as socio-technical systems that require alignment across multiple domains.

The evidence for synergistic effects between different intervention types supports complexity theory approaches to urban sustainability that emphasize non-linear interactions and emergent properties. The finding that integrated approaches often outperform the sum of individual components suggests that urban systems exhibit threshold effects and positive feedbacks that can be leveraged through careful system design.

From an institutional analysis perspective, the implementation barriers documented in this review illustrate the importance of institutional fit between innovations and existing governance structures. The most successful implementations appear to occur where institutional entrepreneurs actively work to align regulatory frameworks, financing mechanisms, and professional practices with sustainable design objectives.

The prominence of co-benefits in justifying and sustaining these interventions supports theoretical frameworks that emphasize the importance of multiple value creation in complex urban systems. Single-objective optimization approaches appear inadequate for addressing the multifaceted challenges of urban sustainability and climate adaptation.

4.3 Practical Implications

The review findings have several important implications for urban planning and design practice. First, the evidence for context-dependent effectiveness suggests that successful implementation requires significant investment in local adaptation and customization. Generic design approaches and standardized solutions are likely to underperform compared to locally adapted strategies.

The importance of integrated approaches suggests that professional practice needs to evolve toward more collaborative and interdisciplinary design processes. The traditional separation between architecture, landscape architecture, and urban planning may be counterproductive for developing effective climate resilience strategies.

The significance of co-benefits for economic justification suggests that project evaluation and financing mechanisms need to incorporate broader impact assessment methods. Traditional cost–benefit analyses that focus only on direct climate impacts likely undervalue these interventions and bias decision-making toward conventional alternatives.

The systematic implementation barriers identified in this review suggest that successful scaling requires coordinated intervention across multiple domains. Technical innovation alone is insufficient; successful widespread implementation requires parallel innovation in regulatory frameworks, financing mechanisms, professional education, and community engagement approaches.

4.4 Policy Implications

The findings have significant implications for urban policy at multiple scales. At the national and regional level, building codes and planning regulations need systematic updating to accommodate and incentivize integrated sustainable design approaches. Model codes and standards could accelerate local adoption by providing tested frameworks that local jurisdictions can adapt.

Tax policy and development incentives represent important levers for encouraging private sector adoption of passive design and nature-based solutions. The evidence for long-term economic benefits suggests that temporary incentives could help overcome market barriers and establish positive precedents that become self-sustaining.

Public investment in green infrastructure can serve as both direct climate adaptation and market development for private sector adoption. Strategic public investments in visible and successful projects can demonstrate feasibility and build community support for broader implementation.

Professional licensing and education requirements represent important policy levers for building implementation capacity. Requirements for continuing education in sustainable design could accelerate knowledge diffusion and improve implementation quality.

Procurement policies for public sector projects can create important market signals and learning opportunities. Requirements or preferences for sustainable design in public projects can help develop local expertise and supply chains that support broader market adoption.

4.5 Applications and Recommendations

Based on the review findings, several specific applications and recommendations emerge for different stakeholder groups (see Table 11):

5 Limitations of the Review

This systematic review has several important limitations that should be considered when interpreting the findings and applying them in practice (see Table 12).

Table 11. Applications and recommendations by stakeholder group

Stakeholder Group	Applications and Recommendations
Urban Planners and Designers	– Prioritize integrated approaches that combine multiple passive design strategies and nature-based solutions rather than single-intervention approaches – Invest in local climate analysis and adaptation to optimize interventions for specific contexts – Develop comprehensive monitoring and evaluation protocols to document performance and build evidence base – Engage communities early and continuously in design processes to build support and ensure long-term success
Policymakers	– Update building codes and zoning regulations to accommodate and incentivize integrated sustainable design – Develop financing mechanisms that capture long-term and co-benefit values of sustainable design investments – Establish performance standards and monitoring requirements for green infrastructure to ensure quality implementation – Create professional development requirements and incentives for sustainable design education
Developers and Property Owners	– Consider lifecycle economic analysis that incorporates energy savings, property value premiums, and other long-term benefits – Invest in high-quality design and implementation to ensure performance and avoid negative precedents – Develop maintenance protocols and budgets appropriate for nature-based solutions and passive design features – Market sustainable design features effectively to capture property value premiums and tenant demand
Communities and Advocacy Groups	– Advocate for equitable distribution of green infrastructure investments to avoid gentrification effects – Participate actively in planning processes to ensure community needs and preferences are incorporated – Support policies that require or incentivize sustainable design in new development and major renovations – Build local capacity for green infrastructure maintenance

6 Future Research Directions

The systematic review identified several important research gaps and opportunities for future investigation that could significantly advance the field of passive design and nature-based solutions for climate-resilient cities (see Table 13).

While Table 13 outlines a broad range of research gaps and future directions, three priorities emerge as particularly critical for advancing both research and practice. First, the lack of integrated and systems-based frameworks for combining

Table 12. Summary of review limitations and their implications

Limitation Area	Description
Language and Scope	– Review limited to peer-reviewed English-language literature, excluding potentially valuable case studies from other languages or gray literature, especially from non-English speaking regions or practice-based organizations
Study Heterogeneity	– Variability in study designs, outcome measures, and contexts prevented quantitative meta-analysis – lack of standardized metrics reduced precision of effect size estimates
Publication Bias	– Positive results more likely published than null/negative findings, leading to possible overestimation of effectiveness and underestimation of challenges
Time Frame	– Review period (2010–2025) excluded earlier foundational work – publication lag may cause underrepresentation of recent innovations and trends
Quality Assessment Bias	– Assessment criteria involved subjective judgments – exclusion of lower-scoring studies may have removed practically valuable lessons despite limited rigor
Contextual Focus	– Emphasis on urban contexts excluded rural and peri-urban studies that might provide additional insights under different conditions
Geographic Bias	– Overrepresentation of studies from developed countries limits generalizability to developing contexts, where urban growth and unique constraints are most critical
Rapid Technological Change	– Findings on cost-effectiveness and technical performance may become outdated quickly due to rapid innovation in design and technologies

passive design strategies and nature-based solutions remains a major barrier to scaling successful interventions beyond pilot projects. Second, insufficient long-term performance and maintenance evidence particularly across diverse climatic and socio-economic contexts continues to undermine confidence among policymakers, investors, and practitioners. Third, persistent gaps in policy and governance research, including regulatory alignment, valuation mechanisms, and multi-level coordination, significantly constrain mainstream adoption. Addressing these priority gaps would provide the greatest leverage for overcoming current implementation barriers and accelerating the transition toward climate-resilient urban development.

Table 13. Identified research gaps and future research directions

Research Area	Key Gaps Identified	Future Research Directions
Long-term Performance and Maintenance	– Lack of long-term monitoring (mostly 1–5 years studied) – Limited knowledge on durability, performance degradation, and lifecycle costs – Inadequate maintenance as major cause of failure	– Establish long-term monitoring across climates – Study degradation patterns & maintenance effectiveness – Develop cost-effective & community-based maintenance models
Integrated Systems	– Focus mainly on individual interventions – Limited research on combined/passive + NBS systems – Lack of systems-level performance data	– Study synergistic effects & optimal configurations – Apply systems thinking to capture non-linear interactions – Investigate optimal scales (building, neighborhood, city)
Social Equity & Environmental Justice	– Risk of gentrification & inequitable benefit distribution – Lack of culturally appropriate approaches	– Assess distributional impacts of green infrastructure – Develop equitable access strategies – Use participatory research to reflect community priorities
Economic Valuation & Financing	– Inconsistent monetization of co-benefits – Lack of standardized ecosystem valuation methods – Limited financing models aligned with benefits	– Develop standardized valuation approaches – Explore innovative financing (green bonds, PES, tax increment financing) – Align financing with long-term benefit profiles
Climate Change Adaptation	– Limited research under future climate scenarios – Focus mostly on current conditions	– Test resilience under extreme heat, rainfall, extreme events – Study adaptive management & modular designs – Explore flexible systems adaptable to changing climate
Developing Country Contexts	– Geographic bias toward developed countries – Lack of locally appropriate/ low-cost approaches	– Research passive & NBS in resource-constrained settings – Study low-cost, locally suitable solutions – Explore synergies with poverty reduction, health, & economic development

(continued)

Table 13. (continued)

Research Area	Key Gaps Identified	Future Research Directions
Technology Integration	– Limited research on sensors, AI, digital tools – Weak integration with traditional passive & NBS methods	– Study integration of smart controls & AI – Assess role of digital platforms in engagement & monitoring – Maintain low-impact characteristics while enhancing performance
Policy & Governance	– Institutional barriers underexplored – Limited comparative policy studies	– Study successful governance & policy innovations – Compare regulatory & incentive structures – Investigate multi-level governance coordination

7 Conclusion

This systematic literature review has provided a comprehensive analysis of passive design strategies and nature-based solutions for climate-resilient cities, synthesizing evidence from 108 high-quality studies published between 2010 and 2025. The findings demonstrate substantial potential for these approaches to address multiple urban climate challenges while providing significant co-benefits for human health, environmental quality, and economic development.

7.1 Integrated Answers to the Research Questions

In response to RQ1, this systematic literature review identified a rich and comprehensive taxonomy of passive design strategies and nature-based solutions implemented across diverse urban contexts. Building-scale passive design strategies including natural ventilation, thermal mass utilization, solar orientation, and adaptive building envelopes demonstrated consistent effectiveness in reducing energy consumption and improving indoor thermal comfort. At the urban scale, interventions related to urban morphology, street design, and public space configuration showed the capacity to generate neighborhood and district-level climate resilience benefits. Nature-based solutions, particularly green infrastructure and water-sensitive urban design, emerged as dominant and widely studied categories, with urban tree canopy, green roofs, and constructed wetlands representing the most frequently implemented interventions.

Addressing RQ2, the empirical evidence demonstrates that the effectiveness of passive design strategies and nature-based solutions varies across climate challenges and implementation contexts. The most pronounced and consistent impacts were observed in urban heat island mitigation, where urban forests, green roofs, and cool surfaces produced measurable and repeatable temperature reductions. Stormwater management through nature-based solutions showed strong and often superior performance compared to conventional gray infrastructure, particularly in runoff retention, peak flow reduction, and water quality improvement. Energy efficiency gains from passive design strategies were substantial, especially when multiple measures were combined in integrated design approaches. However, the magnitude of impacts across all domains remained strongly context-dependent, influenced by climate conditions, urban form, and design quality.

Beyond technical effectiveness, RQ3 highlights that the socio-environmental co-benefits of passive design and nature-based solutions frequently exceed their direct climate adaptation benefits. The reviewed literature consistently documents improvements in public health, thermal comfort, mental well-being, biodiversity conservation, and local economic development. Economic benefits, including energy cost savings, infrastructure lifecycle cost reductions, and property value premiums, further strengthen the case for implementation. At the same time, the review identifies important trade-offs and unintended consequences, such as increased maintenance requirements, water consumption in arid regions, land-use competition, and risks of green gentrification. These findings underscore the necessity of equity-sensitive and context-aware planning frameworks.

With respect to RQ4, the review reveals that systematic implementation barriers continue to limit the widespread adoption of passive and nature-based strategies. Institutional and regulatory misalignment, fragmented governance structures, high upfront capital costs, limited access to appropriate financing mechanisms, and professional knowledge gaps were consistently identified across studies. Social and cultural barriers, including public acceptance challenges and risk aversion among developers, further constrain scaling efforts. These barriers indicate that technical feasibility alone is insufficient; coordinated policy reform, institutional innovation, and capacity building are essential to bridge the gap between research and practice.

Finally, addressing RQ5, the synthesis demonstrates that integrated approaches combining multiple passive design strategies and nature-based solutions consistently outperform isolated interventions. Synergistic configurations yield higher energy savings, stronger climate resilience outcomes, and more robust co-benefits at both building and urban scales. Despite this evidence, the review finds that systematic frameworks for integration remain underdeveloped, and practical guidance for combining strategies across scales is limited. Advancing integrative design methodologies and cross-sector collaboration therefore represents a critical priority for future research and practice.

7.2 Answers to Research Questions

For clarity and reference, the key answers to RQ1–RQ5 are summarized in the following table (see Table 14).

Table 14. Direct answers to research questions

Research Question (RQ)	Findings
RQ1: Categories of Interventions	The review identified comprehensive taxonomies of both passive design strategies (building scale and urban scale) and nature-based solutions (green infrastructure, water-sensitive urban design, and biodiversity-focused interventions). The most extensively implemented and studied approaches include natural ventilation systems, green roofs, urban tree canopy, and constructed wetlands for stormwater management
RQ2: Effectiveness Evidence	Strong empirical evidence demonstrates effectiveness in urban heat island mitigation (0.5–2.0 °C air temperature reductions from urban forests), stormwater management (45–90% retention rates for green roofs), and energy efficiency (20–60% cooling energy reductions from natural ventilation). Effectiveness shows strong context dependencies related to climate, urban morphology, and implementation quality
RQ3: Co-benefits and Trade-offs	Extensive co-benefits include public health improvements, property value premiums (3–15%), biodiversity conservation, and local economic development. Trade-offs include water consumption requirements, maintenance costs (20–40% higher than conventional alternatives), and potential gentrification effects requiring proactive management
RQ4: Implementation Barriers	Systematic barriers include regulatory misalignment, capital cost premiums (10–50% higher upfront costs), professional knowledge gaps, and social acceptance challenges. These barriers require coordinated intervention across multiple domains rather than purely technical solutions
RQ5: Integration and Synergies	Integrated approaches combining multiple strategies demonstrate superior performance compared to individual interventions, with comprehensive passive design achieving 50–80% energy savings. However, frameworks for systematic integration remain underdeveloped and represent a critical research and practice need

The findings have several critical implications for advancing climate-resilient urban development. Policymakers need to prioritize regulatory reform and institutional innovation alongside technical innovation to address the systematic barriers limiting implementation. This includes updating building codes, developing new financing mechanisms, and creating professional development requirements.

Practitioners need to embrace integrated approaches that combine multiple strategies and engage in collaborative, interdisciplinary design processes. The evidence clearly demonstrates that system-level approaches outperform individual interventions, requiring evolution in professional practice toward more holistic and collaborative approaches.

Communities and advocacy groups have important roles in ensuring equitable implementation and building social support for these approaches. The success of passive design and nature-based solutions depends critically on community acceptance and engagement, requiring meaningful participation in planning and design processes.

The urgency of climate change demands rapid scaling of effective solutions, but the evidence from this review suggests that successful scaling requires careful attention to local adaptation, quality implementation, and systematic barrier removal. The challenge for the field is to maintain high implementation standards while achieving the speed and scale of deployment necessary to address climate challenges.

The integration of passive design strategies and nature-based solutions represents a promising pathway toward climate-resilient cities, but realizing this potential requires coordinated effort across multiple domains and scales. This review provides the evidence base and analytical framework to support that coordinated effort, but the ultimate success will depend on the commitment and collaboration of all stakeholders in the urban development process.

References

1. Abanda FH, Byers L (2016) An investigation of the impact of building orientation on energy consumption in a domestic building using emerging BIM (Building Information Modelling). Energy 97:517–527
2. Abhijith KV, Kumar P, Gallagher J, McNabola A, Baldauf R, Pilla F, Pulvirenti B (2017) Air pollution abatement performances of green infrastructure in open road and built-up street canyon environments–A review. Atmos Environ 162:71–86
3. Achour-Younsi S, Kharrat F (2016) Outdoor thermal comfort: Impact of the geometry of an urban street canyon in a Mediterranean subtropical climate–Case study Tunis, Tunisia. Procedia Soc Behav Sci 216:689–700
4. Adams C, Moglia M, Frantzeskaki N (2024) Design principles for mainstreaming of nature-based solutions in cities: A proposal for future pathways. Nature-Based Solutions 6:100155
5. Aflaki A, Mirnezhad M, Ghaffarianhoseini A, Ghaffarianhoseini A, Omrany H, Wang ZH, Akbari H (2015) Urban heat island mitigation strategies: A state-of-the-art review on Kuala Lumpur, Singapore and Hong Kong. Cities 62:131–145
6. Aghaloo K, Sharifi A, Habibzadeh N, Ali T, Chiu YR (2024) How nature-based solutions can enhance urban resilience to flooding and climate change and provide other co-benefits: A systematic review and taxonomy. Urban Forestry & Urban Greening 95:128320
7. Aghaloo K, Sharifi A (2025) Integrated spatial prioritization of urban nature-based solutions for climate adaptation, mitigation, and justice. Int J Sust Dev World 32(2):224–241
8. Ahmad N, Hassan Q (2025) Ecosystem services linked to nature-based solutions for resilient and sustainable cities in India. Frontiers in Water 6:1504492
9. Akeiber H, Nejat P, Majid MZA, Wahid MA, Jomehzadeh F, Zeynali Famileh I, Zaki SA (2016) A review on phase change material (PCM) for sustainable passive cooling in building envelopes. Renew Sustain Energy Rev 60:1470–1497
10. Al-Rubaei AM, Stenglein AL, Viklander M, Blecken GT (2013) Long-term hydraulic performance of porous asphalt pavements in northern Sweden. J Irrig Drain Eng 139(6):499–505
11. Al-Tamimi NA, Fadzil SFS (2011) The potential of shading devices for temperature reduction in high-rise residential buildings in the tropics. Procedia Engineering 21:273–282
12. Albert C, Schröter B, Haase D, Brillinger M, Henze J, Herrmann S, Matzdorf B (2019) Addressing societal challenges through nature-based solutions: How can landscape planning and governance research contribute? Landsc Urban Plan 182:12–21
13. Alves A, Gersonius B, Kapelan Z, Vojinovic Z, Sanchez A (2019) Assessing the co-benefits of green-blue-grey infrastructure for sustainable urban flood risk management. J Environ Manage 239:244–254

14. Andersson E, Haase D, Scheuer S, Wellmann T (2015) Neighbourhood character affects the spatial extent and magnitude of the functional footprint of urban green infrastructure. Landscape Ecol 30(7):1153–1167
15. Anderson V, Gough WA (2022) A typology of nature-based solutions for sustainable development: an analysis of form, function, nomenclature, and associated applications. Land 11(7):1072
16. Anderson V, Suneja M, Dunjic J (2023) Sensing and measurement techniques for evaluation of nature-based solutions: a state-of-the-art review. Land 12(8):1477
17. Anguelovski I, Connolly JJ, Garcia-Lamarca M, Cole H, Pearsall H (2019) New scholarly pathways on green gentrification: What does the urban 'green turn' mean and where is it going? Prog Hum Geogr 43(6):1064–1086
18. Anjum B, Sultana R, Saddaf N (2024) The effectiveness of nature-based solutions to address climate change in Dhaka Bangladesh. Social Sciences & Humanities Open 10:100985
19. Aronson MF, Lepczyk CA, Evans KL, Goddard MA, Lerman SB, MacIvor JS, Vargo T (2017) Biodiversity in the city: key challenges for urban green space management. Front Ecol Environ 15(4):189–196
20. Artmann M, Sartison K (2018) The role of urban agriculture as a nature-based solution: A review for developing a systemic assessment framework. Sustainability 10(6):1937
21. Artmann M, Inostroza L, Fan P (2019) Urban sprawl, compact urban development and green cities. How much do we know, how much do we agree? Ecol Ind 96:3–9
22. Ascione F, Bianco N, De Rossi F, Turni G, Vanoli GP (2013) Green roofs in European climates. Are effective solutions for the energy savings in air-conditioning? Appl Energy 104:845–859
23. Ástmarsson B, Jensen PA, Maslesa E (2013) Sustainable renovation of residential buildings and the landlord/tenant dilemma. Energy Policy 63:355–362
24. Attia S, Lioure R, Declaude Q (2018) Future trends and main concepts of adaptive facade systems. Energy Science & Engineering 8(9):3255–3272
25. Baldock KC, Goddard MA, Hicks DM, Kunin WE, Mitschunas N, Morse H, Memmott J (2019) A systems approach reveals urban pollinator hotspots and conservation opportunities. Nature Ecology & Evolution 3(3):363–373
26. Berndtsson JC (2010) Green roof performance towards management of runoff water quantity and quality: A review. Ecol Eng 36(4):351–360
27. Bhamare DK, Rathod MK, Banerjee J (2019) Passive cooling techniques for building and their applicability in different climatic zones—The state of art. Energy and Buildings 198:467–490
28. Bianchini F, Hewage K (2012) How "green" are the green roofs? Lifecycle analysis of green roof materials. Build Environ 48:57–65
29. Bird S, Hernández D (2012) Policy options for the split incentive: Increasing energy efficiency for low-income renters. Energy Policy 48:506–514
30. Bissett N, Saunders J, Guenther J (2022) Developing place-based climate adaptation pathways with disadvantaged communities: Learning from experience in western Sydney. J Environ Planning Manage 65(2):334–354
31. Bouzouidja R, Cannavo P, Bodénan P, Gulyás Á, Kiss M, Kovács A, Rodriguez F (2021) How to evaluate nature-based solutions performance for microclimate, water and soil management issues–Available tools and methods from Nature4Cities European project results. Ecol Ind 125:107556
32. Brown RD, Vanos J, Kenny N, Lenzholzer S (2015) Designing urban parks that ameliorate the effects of climate change. Landsc Urban Plan 138:118–131
33. Buccolieri R, Santiago JL, Rivas E, Sanchez B (2018) Review on urban tree modelling in CFD simulations: Aerodynamic, deposition and thermal effects. Urban Forestry & Urban Greening 31:212–220
34. Calliari E, Castellari S, Davis M, Linnerooth-Bayer J, Martin J, Mysiak J, Zandersen M (2022) Building climate resilience through nature-based solutions in Europe: A review of enabling knowledge, finance and governance frameworks. Clim Risk Manag 37:100450
35. Callway R, Dixon T, Nikolic D (2019) Embedded green building innovation: An ethnographic study of a property development organisation. Constr Innov 19(4):555–574

36. Campisano A, Butler D, Ward S, Burns MJ, Friedler E, DeBusk K, Han M (2017) Urban rainwater harvesting systems: Research, implementation and future perspectives. Water Res 115:195–209
37. Camps-Calvet M, Langemeyer J, Calvet-Mir L, Gómez-Baggethun E (2016) Ecosystem services provided by urban gardens in Barcelona, Spain: Insights for policy and planning. Environ Sci Policy 62:14–23
38. Cao G, Awbi H, Yao R, Fan Y, Sirén K, Kosonen R, Zhang JJ (2014) A review of the performance of different ventilation and airflow distribution systems in buildings. Build Environ 73:171–186
39. Cariñanos P, Casares-Porcel M (2011) Urban green zones and related pollen allergy: A review. Some guidelines for designing spaces with low allergy impact. Landsc Urban Plan 101(3):205–214
40. Carter JG, Cavan G, Connelly A, Guy S, Handley J, Kazmierczak A (2015) Climate change and the city: Building capacity for urban adaptation. Prog Plan 95:1–66
41. Castelo S, Amado M, Ferreira F (2023) Challenges and opportunities in the use of nature-based solutions for urban adaptation. Sustainability 15(9):7243
42. Castleton HF, Stovin V, Beck SB, Davison JB (2010) Green roofs; building energy savings and the potential for retrofit. Energy and Buildings 42(10):1582–1591
43. Chan FKS, Lu L, Zhu Y, Balzan MV, Pezzoli A, Johnson M, Xu Y (2023) Exploring community perceptions and engagement of nature-based solutions: The case of Ningbo, a Chinese coastal sponge city. Nature-Based Solutions 4:100093
44. Chen Y, Tong Z, Zheng Y, Samuelson H, Norford L (2020) Transfer learning with deep neural networks for model predictive control of HVAC and natural ventilation in smart buildings. J Clean Prod 254:119866
45. Chenari B, Dias Carrilho J, Gameiro da Silva M (2016) Towards sustainable, energy-efficient and healthy ventilation strategies in buildings: A review. Renew Sustain Energy Rev 59:1426–1447
46. Cole HV, Garcia Lamarca M, Connolly JJ, Anguelovski I (2017) Are green cities healthy and equitable? Unpacking the relationship between health, green space and gentrification. J Epidemiol Community Health 71(11):1118–1121
47. Coma J, Pérez G, de Gracia A, Burés S, Urrestarazu M, Cabeza LF (2017) Vertical greenery systems for energy savings in buildings: A comparative study between green walls and green facades. Build Environ 111:228–237
48. Cooper C, Cunningham N, Bracken LJ (2023) Exploring different framings of nature-based solutions with respect to governance, and citizen participation, beneficiaries, and quality of life outcomes. Environ Sci Policy 150:103592
49. Cortinovis C, Geneletti D (2018) Ecosystem services in urban plans: What is there, and what is still needed for better decisions. Land Use Policy 70:298–312
50. Cousins JJ (2021) Justice in nature-based solutions: Research and pathways. Ecol Econ 180:106874
51. Cousins JJ, Newell JP, Steinberg L (2019) (Re) constructing stormwater as a resource: A critical review of green infrastructure representations in the media. Cities 95:102384
52. Croeser T, Garrard G, Sharma R, Ossola A, Bekessy S (2021) Choosing the right nature-based solutions to meet diverse urban challenges. Urban Forestry & Urban Greening 65:127337
53. Darko A, Chan APC, Ameyaw EE, He BJ, Olanipekun AO (2017) Examining issues influencing green building technologies adoption: The United States green building experts' perspectives. Energy and Buildings 144:320–332
54. Davies C, Lafortezza R (2017) Urban green infrastructure in Europe: Is greenspace planning and policy compliant? Land Use Policy 69:93–101
55. Davies ZG, Edmondson JL, Heinemeyer A, Leake JR, Gaston KJ (2011) Mapping an urban ecosystem service: Quantifying above-ground carbon storage at a city-wide scale. J Appl Ecol 48(5):1125–1134
56. Davis AP, Hunt WF, Traver RG, Clar M (2012) Bioretention technology: Overview of current practice and future needs. J Environ Eng 135(3):109–117

57. De Beeck TO, den Heijer C, Coppens T (2024) Financing climate adaptation in Flemish cities: Unpacking financial strategies and policy dynamics for nature-based solutions. Landsc Urban Plan 248:105094
58. Demuzere M, Orru K, Heidrich O, Olazabal E, Geneletti D, Orru H, Faehnle M (2014) Mitigating and adapting to climate change: Multi-functional and multi-scale assessment of green urban infrastructure. J Environ Manage 146:107–115
59. Dhakal KP, Chevalier LR (2017) Managing urban stormwater for urban sustainability: Barriers and policy solutions for green infrastructure application. J Environ Manage 203:171–181
60. Di Pirro E, Sallustio L, Castellar JA, Sgrigna G, Marchetti M, Lasserre B (2022) Facing multiple environmental challenges through maximizing the co-benefits of nature-based solutions at a national scale in Italy. Forests 13(4):548
61. Donatti CI, Andrade A, Cohen-Shacham E, Fedele G, Hou-Jones X, Robyn B (2022) Ensuring that nature-based solutions for climate mitigation address multiple global challenges. One Earth 5(5):493–504
62. Dorst H, Van Der Jagt A, Toxopeus H, Tozer L, Raven R, Runhaar H (2022) What's behind the barriers? Uncovering structural conditions working against urban nature-based solutions. Landsc Urban Plan 220:104335
63. Duffaut C, Frascaria-Lacoste N, Versini PA (2022) Barriers and levers for the implantation of sustainable nature-based solutions in cities: Insights from France. Sustainability 14(16):9975
64. Dzhambov AM, Markevych I, Hartig T, Tilov B, Arabadzhiev Z, Stoyanov D, Dimitrova DD (2018) Multiple pathways link urban green-and bluespace to mental health in young adults. Environ Res 166:223–233
65. Elmqvist T, Setälä H, Handel SN, van der Ploeg S, Aronson J, Blignaut JN, de Groot R (2015) Benefits of restoring ecosystem services in urban areas. Current Opinion in Environmental Sustainability 14:101–108
66. Endreny T, Santagata R, Perna A, De Stefano C, Rallo RF, Ulgiati S (2017) Implementing and managing urban forests: A much needed conservation strategy to increase ecosystem services and urban wellbeing. Ecol Model 360:328–335
67. Everett G, Lamond JE, Morzillo AT, Matsler AM, Chan FKS (2018) Delivering green streets: An exploration of changing perceptions and behaviours over time around bioswales in Portland, Oregon. Journal of Flood Risk Management 11:S973–S985
68. Fernandes CO, da Silva IM, Teixeira CP, Costa L (2019) Between tree lovers and tree haters. Drivers of public perception regarding street trees and its implications on the urban green infrastructure planning. Urban Forestry & Urban Greening 37:97–108
69. Ferrario F, Mourato JM, Rodrigues MS, Dias LF (2024) Evaluating Nature-based Solutions as urban resilience and climate adaptation tools: A meta-analysis of their benefits on heatwaves and floods. Sci Total Environ 950:175179
70. Ferreira V, Barreira AP, Loures L, Antunes D, Panagopoulos T (2020) Stakeholders' engagement on nature-based solutions: A systematic literature review. Sustainability 12(2):640
71. Flynn KM, Traver RG (2013) Green infrastructure life cycle assessment: A bio-infiltration case study. Ecol Eng 55:9–22
72. Fowdar HS, Hatt BE, Breen P, Cook PL, Deletic A (2017) Designing living walls for greywater treatment. Water Res 110:218–232
73. Frantzeskaki N (2019) Seven lessons for planning nature-based solutions in cities. Environ Sci Policy 93:101–111
74. Gao Y, Xu J, Yang S, Tang X, Zhou Q, Ge J, Levinson R (2014) Cool roofs in China: Policy review, building simulations, and proof-of-concept experiments. Energy Policy 74:190–214
75. Garrett JK, White MP, Huang J, Ng S, Hui Z, Leung C, Wong MC (2019) Urban blue space and health and wellbeing in Hong Kong: Results from a survey of older adults. Health Place 55:100–110
76. Gascon M, Triguero-Mas M, Martínez D, Dadvand P, Forns J, Plasència A, Nieuwenhuijsen MJ (2015) Mental health benefits of long-term exposure to residential green and blue spaces: A systematic review. Int J Environ Res Public Health 12(4):4354–4379

77. Gascon M, Triguero-Mas M, Martínez D, Dadvand P, Rojas-Rueda D, Plasència A, Nieuwenhuijsen MJ (2016) Residential green spaces and mortality: A systematic review. Environ Int 86:60–67
78. Geetha NB, Velraj R (2012) Passive cooling methods for energy efficient buildings with and without thermal energy storage–A review. Energy Education Science and Technology Part A: Energy Science and Research 29(2):913–946
79. Geneletti D, Zardo L (2016) Ecosystem-based adaptation in cities: An analysis of European urban climate adaptation plans. Land Use Policy 50:38–47
80. Ghofrani Z, Sposito V, Faggian R (2017) A comprehensive review of blue-green infrastructure concepts. International Journal of Environment and Sustainability 6(1):15–36
81. Gil-Baez M, Barrios-Padura Á, Molina-Huelva M, Chacartegui R (2019) Natural ventilation systems in 21st-century for near zero energy school buildings. Energy 137:1186–1200
82. Golden HE, Hoghooghi N (2018) Green infrastructure and its catchment-scale effects: An emerging science. Wiley Interdiscip Rev Water 5(1):e1254
83. Grafius DR, Corstanje R, Warren PH, Evans KL, Hancock S, Harris JA (2016) The impact of land use/land cover scale on modelling urban ecosystem services. Landscape Ecol 31(7):1509–1522
84. Gromke C, Blocken B (2015) Influence of avenue-trees on air quality at the urban neighborhood scale. Part I: Quality assurance studies and turbulent Schmidt number analysis for RANS CFD simulations. Environ Pollut 196:214–223
85. Guerra-Santin O, Silvester S (2017) Development of Dutch occupancy and heating profiles for building simulation. Building Research & Information 45(4):396–413
86. Haaland C, van den Bosch CK (2015) Challenges and strategies for urban green-space planning in cities undergoing densification: A review. Urban Forestry & Urban Greening 14(4):760–771
87. Haase D, Larondelle N, Andersson E, Artmann M, Borgström S, Breuste J, Kabisch N (2014) A quantitative review of urban ecosystem service assessments: Concepts, models, and implementation. Ambio 43(4):413–433
88. Hall DM, Camilo GR, Tonietto RK, Ollerton J, Ahrné K, Arduser M, Threlfall CG (2017) The city as a refuge for insect pollinators. Conserv Biol 31(1):24–29
89. Han Y, Zhou Z, Wang C, Li W, Liu Y, Jia Z (2025) Arranging the inherent thermal mass of urban buildings enhances passive energy harvesting in winter. Energy, 137482
90. Hand KL, Freeman C, Seddon PJ, Recio MR, Stein A, van Heezik Y (2017) The importance of urban gardens in supporting children's biophilia. Proc Natl Acad Sci 114(2):274–279
91. Hansen R, Olafsson AS, van der Jagt AP, Rall E, Pauleit S (2019) Planning multifunctional green infrastructure for compact cities: What is the state of practice? Ecol Ind 96:99–110
92. Haque MT, Geronimo FKF, Robles MEL, Vispo C, Kim LH (2025) Comparative evaluation of the carbon storage capacities in urban stormwater nature-based technologies. Ecol Eng 212:107539
93. Harvey LD (2013) Recent advances in sustainable buildings: Review of the energy and cost performance of the state-of-the-art best practices from around the world. Annu Rev Environ Resour 38:281–309
94. Hassan AM, ELMokadem AA, Megahed NA, Eleinen OMA (2020) Urban morphology as a passive strategy in promoting outdoor air quality. Journal of Building Engineering 29:101204
95. Heaviside C, Macintyre H, Vardoulakis S (2017) The urban heat island: Implications for health in a changing environment. Current Environmental Health Reports 4(3):296–305
96. Heffernan E, Pan W, Liang X, De Wilde P (2015) Zero carbon homes: Perceptions from the UK construction industry. Energy Policy 79:23–36
97. Hegazy B, Khodeir L, Fathy, F (2025) Achieving socio-economic resilience in neighborhood through nature-based solutions: A systematic review. Results Eng, 104266.
98. Hondula DM, Balling RC, Vanos JK, Georgescu M (2015) Rising temperatures, human health, and the role of adaptation. Current Climate Change Reports 1(3):144–154
99. Hoyle H, Hitchmough J, Jorgensen A (2017) All about the 'wow factor'? The relationships between aesthetics, restorative effect and perceived biodiversity in designed urban planting. Landsc Urban Plan 164:109–123

100. Hunter RF, Cleland C, Cleary A, Droomers M, Wheeler BW, Sinnett D, Braubach M (2019) Environmental, health, wellbeing, social and equity effects of urban green space interventions: A meta-narrative evidence synthesis. Environ Int 130:104923
101. Hwang BG, Shan M, Supa'at NNB (2017) Green commercial building projects in Singapore: Critical risk factors and mitigation measures. Sustain Cities Soc 30:237–247
102. Ihara T, Kikegawa Y, Asahi K, Genchi Y, Kondo H (2015) Changes in year-round air temperature and annual energy consumption in office building areas by urban heat-island countermeasures and energy-saving measures. Appl Energy 52:273–280
103. Jamali B, Bach PM, Deletic A (2020) Rainwater harvesting for urban flood management–An integrated modelling framework. Water Res 171:115372
104. James P, Banay RF, Hart JE, Laden F (2015) A review of the health benefits of greenness. Current Epidemiology Reports 2(2):131–142
105. Jamei E, Rajagopalan P, Seyedmahmoudian M, Jamei Y (2016) Review on the impact of urban geometry and pedestrian level greening on outdoor thermal comfort. Renew Sustain Energy Rev 54:1002–1017
106. Jay O, Capon A, Berry P, Broderick C, de Dear R, Havenith G, Ebi KL (2021) Reducing the health effects of hot weather and heat extremes: From personal cooling strategies to green cities. The Lancet 398(10301):709–724
107. Jayasooriya VM, Ng AWM (2014) Tools for modeling of stormwater management and economics of green infrastructure practices: A review. Water Air Soil Pollut 225(8):1–20
108. Jayasooriya VM, Ng AWM, Muthukumaran S, Perera BJC (2017) Green infrastructure practices for improvement of urban air quality. Urban Forestry & Urban Greening 21:34–47
109. Jia H, Yao H, Tang Y, Yu SL, Zhen JX, Lu Y (2015) Development of a multi-criteria index ranking system for urban runoff best management practices (BMPs) selection. Environ Monit Assess 185(9):7915–7933
110. Jiménez-Ariza SL, Rey CV, Rodríguez JP, Guzmán-Ramírez, M (2023) Multi-criteria decision analysis inputs for planning the implementation of nature-based solutions in urban contexts. ACE: Architecture, City and Environment 18(52)
111. Johannessen, Å, Wamsler, C (2017) What does resilience mean for urban water services? Ecol Soc 22(1)
112. Jomehzadeh F, Nejat P, Calautit JK, Yusof MBM, Zaki SA, Hughes BR, Yazid MNAWM (2017) A review on windcatcher for passive cooling and natural ventilation in buildings, Part 1: Indoor air quality and thermal comfort assessment. Renew Sustain Energy Rev 70:736–756
113. Kabisch N, Frantzeskaki N, Pauleit S, Naumann S, Davis M, Artmann M, Bonn, A (2016) Nature-based solutions to climate change mitigation and adaptation in urban areas: Perspectives on indicators, knowledge gaps, barriers, and opportunities for action. Ecol Soc 21(2)
114. Kamali M, Delkash M, Tajrishy M (2017) Evaluation of permeable pavement responses to urban surface runoff. J Environ Manage 187:43–53
115. Keeler BL, Hamel P, McPhearson T, Hamann MH, Donahue ML, Meza Prado KA, Wood SA (2019) Social-ecological and technological factors moderate the value of urban nature. Nature Sustainability 2(1):29–38
116. Khan MOD, Keesstra S, Słowik-Opoka E, Klamerus-Iwan A, Liaqat, W (2025) Determining the role of urban Greenery in soil hydrology: A bibliometric analysis of nature-based solutions in urban ecosystem. Water 17(3): 322
117. Kia A, Wong HS, Cheeseman CR (2021) Defining clogging potential for permeable concrete. J Environ Manage 220:44–53
118. Kim HW, Miller JT (2019) The economic value of green infrastructure in urban areas: A hedonic approach. Landsc Urban Plan 185:245–255
119. Kirimtat A, Koyunbaba BK, Chatzikonstantinou I, Sariyildiz S (2016) Review of simulation modeling for shading devices in buildings. Renew Sustain Energy Rev 53:23–49
120. Kitsopoulou A, Bellos E, Tzivanidis C (2024) An up-to-date review of passive building envelope technologies for sustainable design. Energies 17(16):4039

121. Kneifel J (2010) Life-cycle carbon and cost analysis of energy efficiency measures in new commercial buildings. Energy and Buildings 42(3):333–340
122. Ko Y (2018) Trees and vegetation for residential energy conservation: A critical review for evidence-based urban greening in North America. Urban Forestry & Urban Greening 34:318–335
123. Kolaitis DI, Malliotakis E, Kontogeorgos DA, Mandilaras I, Katsourinis DI, Founti MA (2013) Comparative assessment of internal and external thermal insulation systems for energy efficient retrofitting of residential buildings. Energy and Buildings 64:123–131
124. Kolokotsa D, Santamouris M, Zerefos SC (2013) Green and cool roofs' urban heat island mitigation potential in European climates for office buildings under free floating conditions. Sol Energy 95:118–130
125. Kontokosta CE (2016) Modeling the energy retrofit decision in commercial office buildings. Energy and Buildings 131:1–20
126. Kumar P, Druckman A, Gallagher J, Gatersleben B, Allison S, Eisenman TS, Abhijith KV (2019) The nexus between air pollution, green infrastructure and human health. Environ Int 133:105181
127. Kuronuma T, Watanabe H, Ishihara T, Kou D, Toushima K, Ando M, Shindo S (2018) CO2 payoff of extensive green roofs with different vegetation species. Sustainability 10(7):2256
128. LaPoint S, Balkenhol N, Hale J, Sadler J, van der Ree R (2015) Ecological connectivity research in urban areas. Funct Ecol 29(7):868–878
129. Larson KL, Nelson KC, Samples SR, Hall SJ, Bettez N, Cavender-Bares J, Trammell TL (2016) Ecosystem services in managing residential landscapes: Priorities, value dimensions, and cross-regional patterns. Urban Ecosystems 19(1):95–113
130. Lee H, Mayer H, Chen L (2016) Contribution of trees and grasslands to the mitigation of human heat stress in a residential district of Freiburg, Southwest Germany. Landsc Urban Plan 148:37–50
131. LeFevre GH, Paus KH, Natarajan P, Gulliver JS, Novak PJ, Hozalski RM (2015) Review of dissolved pollutants in urban storm water and their removal and fate in bioretention cells. J Environ Eng 141(1):04014050
132. Lepczyk CA, Aronson MF, Evans KL, Goddard MA, Lerman SB, MacIvor JS (2017) Biodiversity in the city: Fundamental questions for understanding the ecology of urban green spaces for biodiversity conservation. Bioscience 67(9):799–807
133. Li D, Bou-Zeid E, Oppenheimer M (2014) The effectiveness of cool and green roofs as urban heat island mitigation strategies. Environ Res Lett 9(5):055002
134. Li X, Zhou Y, Yu S, Jia G, Li H, Li W (2019) Urban heat island impacts on building energy consumption: A review of approaches and findings. Energy 174:407–419
135. Lin BB, Philpott SM, Jha S (2015) The future of urban agriculture and biodiversity-ecosystem services: Challenges and next steps. Basic Appl Ecol 16(3):189–201
136. Liu W, Chen W, Peng C (2016) Influences of setting sizes and combination of green infrastructures on community's stormwater runoff reduction. Ecol Model 318:236–244
137. Liu W, Wei Q, Chen W, Hauer RJ, Gao X (2019) Relationships between green roof plant survival and local climate. Ecol Eng 139:105581
138. Locatelli L, Guerrero M, Russo B, Martínez-Gomariz E, Sunyer D, Martínez M (2020) Socio-economic assessment of green infrastructure for climate change adaptation in the context of urban drainage planning. Sustainability 12(9):3792
139. Lõhmus M, Balbus J (2015) Making green infrastructure healthier infrastructure. Infection Ecology & Epidemiology 5(1):30082
140. Longato D, Cortinovis C, Balzan M, Geneletti D (2024) Identifying suitable policy instruments to promote nature-based solutions in urban plans. Cities 154:105348
141. Lovasi GS, O'Neil-Dunne JP, Lu JW, Sheehan D, Perzanowski MS, MacFaden SW, Rundle A (2013) Urban tree canopy and asthma, wheeze, rhinitis, and allergic sensitization to tree pollen in a New York City birth cohort. Environ Health Perspect 121(4):494–500
142. Lucke T, Beecham S (2011) Field investigation of clogging in a permeable pavement system. Building Research & Information 39(6):603–615

143. Mabon L, Barkved L, de Bruin K, Shih WY (2022) Whose knowledge counts in nature-based solutions? Understanding epistemic justice for nature-based solutions through a multi-city comparison across Europe and Asia. Environ Sci Policy 136:652–664
144. Maksymiuk G, Adamczyk J, Giedych R, Pusłowska-Tyszewska D, Kuchcik M, Cieszewska A (2025) Harnessing street Canyons for comprehensive nature-based solutions. Land (2012), 14(3)
145. Manso M, Castro-Gomes J (2015) Green wall systems: A review of their characteristics. Renew Sustain Energy Rev 41:863–871
146. Markevych I, Schoierer J, Hartig T, Chudnovsky A, Hystad P, Dzhambov AM, Fuertes E (2017) Exploring pathways linking greenspace to health: Theoretical and methodological guidance. Environ Res 158:301–317
147. Martek I, Hosseini MR, Shrestha A, Edwards DJ, Durdyev S (2019) Barriers inhibiting the transition to sustainability within the Australian construction industry: An investigation of technical and social interactions. J Clean Prod 211:281–292
148. Matthews T, Lo AY, Byrne JA (2015) Reconceptualizing green infrastructure for climate change adaptation: Barriers to adoption and drivers for uptake by spatial planners. Landsc Urban Plan 138:155–163
149. McGrane SJ (2016) Impacts of urbanisation on hydrological and water quality dynamics, and urban water management: A review. Hydrol Sci J 61(13):2295–2311
150. McPhearson T, Andersson E, Elmqvist T, Frantzeskaki N (2018) Resilience of and through urban ecosystem services. Ecosyst Serv 12:152–156
151. Meerow S, Newell JP (2017) Spatial planning for multifunctional green infrastructure: Growing resilience in Detroit. Landsc Urban Plan 159:62–75
152. Mell IC, Henneberry J, Hehl-Lange S, Keskin B (2013) Promoting urban greening: Valuing the development of green infrastructure investments in the urban core of Manchester UK. Urban Forestry & Urban Greening 12(3):296–306
153. Mell IC, Henneberry J, Hehl-Lange S, Keskin B (2016) To green or not to green: Establishing the economic value of green infrastructure investments in The Wicker, Sheffield. Urban Forestry & Urban Greening 18:257–267
154. Meraj G, Hashimoto, S (2025) Bridging the adaptation finance gap: the role of nature-based solutions for climate resilience. Sustain Sci, 1–15
155. Middel A, Häb K, Brazel AJ, Martin CA, Guhathakurta S (2014) Impact of urban form and design on mid-afternoon microclimate in Phoenix Local Climate Zones. Landsc Urban Plan 122:16–28
156. Miller SM, Montalto FA (2019) Stakeholder perceptions of the ecosystem services provided by Green Infrastructure in New York City. Ecosyst Serv 37:100928
157. Mirsafa M, Castaldo AG, Lemes de Oliveira F (2025) Enabling nature-based solutions for climate adaptation in cities of the Global South: planning dimensions and cross-cutting pathways for implementation. Environ Dev Sustain 1–25
158. Moh TSY, Jin JJY, Wong LA, Tiong MC, Chan CK (2023) Wind-induced evaporative cooling passive system for tropical hot and humid climate. Frontiers in Mechanical Engineering 9:1069806
159. Moore T, Horne R, Morrissey J (2016) Zero emission housing: Policy development in Australia and comparisons with the EU, UK, USA and California. Environ Innov Soc Trans 11:25–45
160. Morakinyo TE, Dahanayake KKC, Ng E, Chow CL (2017) Temperature and cooling demand reduction by green-roof types in different climates and urban densities: A co-simulation parametric study. Energy and Buildings 145:226–237
161. Morrissey J, Moore T, Horne RE (2011) Affordable passive solar design in a temperate climate: An experiment in residential building orientation. Renewable Energy 36(2):568–577
162. Muerdter CP, Wong CK, LeFevre GH (2018) Emerging investigator series: The role of vegetation in bioretention for stormwater treatment in the built environment: Pollutant removal, hydrologic function, and ancillary benefits. Environ Sci: Water Res Technol 4(5):592–612
163. Müller N, Kuttler W (2021) Counteracting urban climate change: Adaptation measures and their effect on thermal comfort. Theoret Appl Climatol 115(1–2):243–257

164. Niu H, Clark C, Zhou J, Adriaens P (2010) Scaling of economic benefits from green roof implementation in Washington DC. Environ Sci Technol 44(11):4302–4308
165. Norton BA, Coutts AM, Livesley SJ, Harris RJ, Hunter AM, Williams NS (2015) Planning for cooler cities: A framework to prioritise green infrastructure to mitigate high temperatures in urban landscapes. Landsc Urban Plan 134:127–138
166. Nowak DJ, Greenfield EJ, Hoehn RE, Lapoint E (2013) Carbon storage and sequestration by trees in urban and community areas of the United States. Environ Pollut 178:229–236
167. Nowak DJ, Hirabayashi S, Bodine A, Greenfield E (2014) Tree and forest effects on air quality and human health in the United States. Environ Pollut 193:119–129
168. O'Donnell EC, Lamond JE, Thorne CR (2017) Recognising barriers to implementation of Blue-Green Infrastructure: a Newcastle case study. Urban Water Journal 14(9):964–971
169. O'Sullivan OS, Holt AR, Warren PH, Evans KL (2017) Optimising UK urban road verge contributions to biodiversity and ecosystem services with cost-effective management. J Environ Manage 191:162–171
170. Omrani S, Garcia-Hansen V, Capra B, Drogemuller R (2017) Natural ventilation in multi-storey buildings: design process and review of evaluation tools. Build Environ 116:182–194
171. Pacheco-Torres R, Jadraque E, Roldán-Fontana J, Ordóñez J (2015) Analysis of CO2 emissions in the construction phase of single-family detached houses. Sustain Cities Soc 12:63–68
172. Pappalardo V, La Rosa D, Campisano A, La Greca P (2017) The potential of green infrastructure application in urban runoff control for land use planning: A preliminary evaluation from a southern Italy case study. Ecosyst Serv 26:345–354
173. Pataki DE, Alberti M, Cadenasso ML, Felson AJ, McDonnell MJ, Pincetl S, Whitlow TH (2021) The benefits and limits of urban tree planting for environmental and human health. Front Ecol Evol 9:603757
174. Payne EG, Pham T, Deletic A, Hatt BE, Cook PL, Fletcher TD (2019) Which species? A decision-support tool to guide plant selection in stormwater biofilters. Adv Water Resour 113:86–99
175. Peng LL, Jim CY (2013) Green-roof effects on neighborhood microclimate and human thermal sensation. Energies 6(2):598–618
176. Pennino MJ, McDonald RI, Jaffe PR (2016) Watershed-scale impacts of stormwater green infrastructure on hydrology, nutrient fluxes, and combined sewer overflows in the mid-Atlantic region. Sci Total Environ 565:1044–1053
177. Perini K, Rosasco P (2013) Cost–benefit analysis for green façades and living wall systems. Build Environ 70:110–121
178. Pincetl S, Prabhu SS, Gillespie TW, Jenerette GD, Pataki DE (2013) The evolution of tree nursery offerings in Los Angeles County over the last 110 years. Landsc Urban Plan 118:10–17
179. Pisello AL (2017) State of the art on the development of cool coatings for buildings and cities. Sol Energy 144:660–680
180. Prudencio L, Null SE (2018) Stormwater management and ecosystem services: A review. Environ Res Lett 13(3):033002
181. Qaid A, Bin Lamit H, Ossen DR, Raja Shahminan RN (2016) Urban heat island and thermal comfort conditions at micro-climate scale in a tropical planned city. Energy and Buildings 133:577–595
182. Qiao XJ, Kristoffersson A, Randrup TB (2020) Challenges to implementing green infrastructure for urban stormwater management in Cleveland, Ohio. J Am Plann Assoc 84(4):336–349
183. Qin Y (2015) A review on the development of cool pavements to mitigate urban heat island effect. Renew Sustain Energy Rev 52:445–459
184. Radcliffe JC (2019) History of water sensitive urban design/low impact development adoption in Australia and internationally. In Approaches to water sensitive urban design (pp 1–24). Woodhead Publishing
185. Radu G, Chevereşan MI, Perju S, Bărbulescu A (2025) Integrating nature-based solutions for increased resilience to urban flooding in the climate change context. Hydrology 12(1):16

186. Rahman MA, Stratopoulos LM, Moser-Reischl A, Zölch T, Häberle KH, Rötzer T, Pauleit S (2020) Traits of trees for cooling urban heat islands: A meta-analysis. Build Environ 170:106606
187. Ramyar R, Ackerman A, Johnston DM (2021) Adapting cities for climate change through green infrastructure planning. Cities 117:103316
188. Reilly A, Kinnane O (2017) The impact of thermal mass on building energy consumption. Appl Energy 198:108–121
189. Remme RP, Meacham M, Pellowe KE, Andersson E, Guerry AD, Janke B, van Oudenhoven AP (2024) Aligning nature-based solutions with ecosystem services in the urban century. Ecosyst Serv 66:101610
190. Reyes R, Bustamante W, Gironás J, Pastén PA, Rojas V, Suárez F, Moya RE (2016) Effect of substrate depth and roof layers on green roof temperature and water requirements in a semi-arid climate. Ecol Eng 97:624–632
191. Richards DR, Passy P, Oh RR (2017) Impacts of population density and wealth on the quantity and structure of urban green space in tropical Southeast Asia. Landsc Urban Plan 157:553–560
192. Rigolon A, Németh J (2018) We're not in the business of housing:" Environmental gentrification and the nonprofitization of green infrastructure projects. Cities 81:71–80
193. Rizzo A, Bresciani R, Masi F, Boano F, Revelli R, Ridolfi L (2021) Flood reduction as an ecosystem service of constructed wetlands for combined sewer overflow. J Hydrol 560:150–159
194. Roman KK, O'Brien T, Alvey JB, Woo O (2016) Simulating the effects of cool roof and PCM (phase change materials) based roof to mitigate UHI (urban heat island) in prominent US cities. Energy 96:103–117
195. Russo A, Escobedo FJ, Cirella GT, Zerbe S (2016) Edible green infrastructure: an approach and review of provisioning ecosystem services and disservices in urban environments. Agr Ecosyst Environ 242:53–66
196. Saafi K, Daouas N (2018) A life-cycle cost analysis for an optimum combination of cool coating and thermal insulation of residential building roofs in Tunisia. Energy 152:925–938
197. Sádaba J, Luzarraga A, Lenzi S (2025) Designing for climate adaptation: A case study integrating nature-based solutions with urban infrastructure. Urban Science 9(3):74
198. Salameh M, Touqan B (2022) Traditional passive design solutions as a key factor for sustainable modern urban designs in the hot, arid climate of the United Arab Emirates. Buildings 12(11):1811
199. Salvati A, Coch H, Morganti M (2019) Effects of urban compactness on the building energy performance in Mediterranean climate. Energy Procedia 122:499–504
200. Santamouris M (2014) Cooling the cities–a review of reflective and green roof mitigation technologies to fight heat island and improve comfort in urban environments. Sol Energy 103:682–703
201. Santiago JL, Buccolieri R, Rivas E, Calvete-Sogo H, Sanchez B, Martilli A, Martin F (2017) CFD modelling of vegetation barrier effects on the reduction of traffic-related pollutant concentration in an avenue of Pamplona Spain. Sustain Cities Soc 48:101559
202. Sarabi SE, Han Q, Romme AGL, de Vries B, Wendling L (2019) Key enablers of and barriers to the uptake and implementation of nature-based solutions in urban settings: a review. Resources 8(3):121
203. Sarabi S, Han Q, Romme AGL, De Vries B, Valkenburg R, Den Ouden E (2020) Uptake and implementation of nature-based solutions: an analysis of barriers using interpretive structural modeling. J Environ Manage 270:110749
204. Schulze T, Eicker U (2013) Controlled natural ventilation for energy efficient buildings. Energy and Buildings 56:221–232
205. Selmi W, Weber C, Rivière E, Blond N, Mehdi L, Nowak D (2016) Air pollution removal by trees in public green spaces in Strasbourg city, France. Urban Forestry & Urban Greening 17:192–201
206. Şen E, Şat B (2024) Bioswales as engineering functions of nature-based solutions to increase urban resilience: a review. Forestist 74(2):253–258

207. Sera F, Armstrong B, Tobias A, Vicedo-Cabrera AM, Åström C, Bell ML, Gasparrini A (2019) How urban characteristics affect vulnerability to heat and cold: A multi-country analysis. Int J Epidemiol 48(4):1101–1112
208. Shafique M, Kim R, Kyung-Ho K (2018) Green roof for stormwater management in a highly urbanized area: The case of Seoul, Korea. Sustainability 10(3):584
209. Shafique M, Kim R, Rafiq M (2016) Green roof benefits, opportunities and challenges—a review. Renew Sustain Energy Rev 90:757–773
210. Shi L, Chu E, Anguelovski I, Aylett A, Debats J, Goh K, VanDeveer SD (2016) Roadmap towards justice in urban climate adaptation research. Nat Clim Chang 6(2):131–137
211. Skelhorn C, Lindley S, Levermore G (2014) The impact of vegetation types on air and surface temperatures in a temperate city: a fine scale assessment in Manchester, UK. Landsc Urban Plan 121:129–140
212. Song XP, Tan PY, Edwards P, Richards D (2018) The economic benefits and costs of trees in urban forest stewardship: A systematic review. Urban Forestry & Urban Greening 29:162–170
213. Souayfane F, Fardoun F, Biwole PH (2016) Phase change materials (PCM) for cooling applications in buildings: a review. Energy and Buildings 129:396–431
214. Staddon C, Ward S, De Vito L, Zuniga-Teran A, Gerlak AK, Schoeman Y, Booth G (2018) Contributions of green infrastructure to enhancing urban resilience. Environment Systems and Decisions 38(3):330–338
215. Stewart ID, Oke TR (2012) Local climate zones for urban temperature studies. Bull Am Meteor Soc 93(12):1879–1900
216. Stijnen C, Frantzeskaki N, Wijsman K (2024) Beating around the bush: a scoping review of trade-offs for just planning and governance of urban nature-based solutions. Urban Forestry & Urban Greening 102:128525
217. Stovin V, Vesuviano G, De-Ville S (2017) Defining green roof detention performance. Urban Water Journal 14(6):574–588
218. Tan PY, Wong NH, Tan CL, Jusuf SK, Chang MF, Chiam ZQ (2016) A method to partition the relative effects of evaporative cooling and shading on air temperature within vegetation canopy. J Urban Ecol 4(1): juy012
219. Tang Y, Chen A, Zhao S (2016) Carbon storage and sequestration of urban street trees in Beijing China. Front Ecol Evol 4:53
220. Tayouga SJ, Gagné SA (2016) The socio-ecological factors that influence the adoption of green infrastructure. Sustainability 8(12):1277
221. Threlfall CG, Walker K, Williams NS, Hahs AK, Mata L, Stork N, Livesley SJ (2016) The conservation value of urban green space habitats for Australian native bee communities. Biol Cons 187:240–248
222. Tong Z, Chen Y, Malkawi A, Liu Z, Freeman RB (2016) Energy saving potential of natural ventilation in China: The impact of ambient air pollution. Appl Energy 179:660–668
223. Toxopeus H, Polzin F (2021) Reviewing financing barriers and strategies for urban nature-based solutions. J Environ Manage 289:112371
224. Tuhus-Dubrow D, Krarti M (2010) Genetic-algorithm based approach to optimize building envelope design for residential buildings. Build Environ 45(7):1574–1581
225. Valladares-Rendón LG, Schmid G, Lo SL (2017) Review on energy savings by solar control techniques and optimal building orientation for the strategic placement of façade shading systems. Energy and Buildings 140:458–479
226. Valois L, Brachet A, Schiopu N, Barot S (2025) Development of an integrative approach for the assessment of ecosystem services and environmental impacts of urban nature-based solutions. Total Environ Adv, 200133.
227. Vasiliev D, Hazlett RW (2025) Envisaging nature-based solutions as designed ecosystems in the changing world. Land Use Policy 150:107438
228. Velasco E, Roth M, Norford L, Molina LT (2016) Does urban vegetation enhance carbon sequestration? Landsc Urban Plan 148:99–107
229. Venkataramanan V, Lopez D, McCuskey DJ, Kiefus D, McDonald RI, Miller WM, Young SL (2020) Knowledge, attitudes, intentions, and behavior related to green infrastructure for flood management: A systematic literature review. Sci Total Environ 720:137606

230. Versini PA, Ramier D, Berthier E, De Gouvello B (2015) Assessment of the hydrological impacts of green roof: From building scale to basin scale. J Hydrol 524:562–575
231. Viecco M, Vera S, Jorquera H, Bustamante W, Gironás J, Dobbs C, Leiva E (2018) Green roofs and green walls layouts for improved urban air quality by mitigating particulate matter. Build Environ 204:108120
232. Vijayaraghavan K, Biswal BK, Adam MG, Soh SH, Tsen-Tieng DL, Davis AP, Balasubramanian R (2021) Bioretention systems for stormwater management: Recent advances and future prospects. J Environ Manage 292:112766
233. Voskamp IM, Van de Ven FH (2015) Planning support system for climate adaptation: Composing effective sets of blue-green measures to reduce urban vulnerability to extreme weather events. Build Environ 83:159–167
234. Vymazal J (2018) Does clogging affect long-term removal of organics and suspended solids in gravel-based horizontal subsurface flow constructed wetlands? Chem Eng J 331:663–674
235. Wamsler C, Alkan-Olsson J, Björn H, Falck H, Hanson H, Oskarsson T, Zelmerlow F (2020) Beyond participation: When citizen engagement leads to undesirable outcomes for nature-based solutions and climate change adaptation. Clim Change 158(2):235–254
236. Wang J, Banzhaf E, Jones L (2020) Cultural ecosystem services and disservices of urban green infrastructure: A systematic review. J Clean Prod 257:120420
237. Wang Y, Berardi U, Akbari H (2016) Comparing the effects of urban heat island mitigation strategies for Toronto, Canada. Energy and Buildings 114:2–19
238. Wang Y, Ni Z, Chen S, Xia B (2018) Microclimate regulation and energy saving potential from different urban green infrastructures in a subtropical city. J Clean Prod 226:913–927
239. Wang C, Jin J, Davies C, Chen WY (2024) Urban forests as nature-based solutions: a comprehensive overview of the national forest city action in China. Current Forestry Reports 10(2):119–132
240. Weerakkody U, Dover JW, Mitchell P, Reiling K (2018) Evaluating the impact of individual leaf traits on atmospheric particulate matter accumulation using natural and synthetic leaves. Urban Forestry & Urban Greening 30:98–107
241. William R, Goodwell A, Richardson M, Le PV, Kumar P, Stillwell AS (2019) An environmental cost-benefit analysis of alternative green roofing strategies. Ecol Eng 95:1–9
242. Willis KJ, Petrokofsky G (2017) The natural capital of city trees. Science 356(6336):374–376
243. Winston RJ, Dorsey JD, Hunt WF (2016) Quantifying volume reduction and peak flow mitigation for three bioretention cells in clay soils in northeast Ohio. Sci Total Environ 553:83–95
244. Wolch JR, Byrne J, Newell JP (2014) Urban green space, public health, and environmental justice: The challenge of making cities 'just green enough.' Landsc Urban Plan 125:234–244
245. Wolf KL, Lam ST, McKeen JK, Richardson GR, van den Bosch M, Bardekjian AC (2020) Urban trees and human health: A scoping review. Int J Environ Res Public Health 17(12):4371
246. Wolff E, Natakun B, Marome W, Chew G, Seng K, Hamel, P (2025) Nature for the people, by the people: negotiating values, attitudes and behaviours for implementing urban nature-based solutions in social housing. Urban For Urban Green, 128973
247. Xu J, Gao Y, Yang L (2025) Climate-adaptive passive design strategies for near-zero-energy office buildings in central and Southern Anhui, China. Sustainability 17(14):6535
248. Xu X, Liu S, Sun S, Zhang W, Liu Y, Lao Z, Jiang Z (2020) Evaluation of energy saving potential of an urban green space and its water bodies. Energy Build 188:58–70
249. Yang J, Bou-Zeid E (2018) Should cities embrace their heat islands as shields from extreme cold? J Appl Meteorol Climatol 57(6):1309–1320
250. Yang L, Yan H, Lam JC (2014) Thermal comfort and building energy consumption implications–a review. Appl Energy 115:164–173
251. Zhang DL, Shou YX, Dickerson RR (2021) Upstream urbanization exacerbates urban heat island effects. Geophys Res Lett 36(24)
252. Zhang C, Xi C, Feng Z, Wang J, Cao SJ (2022) Passive design for green buildings by using green glass space and earth air tunnel. Energy and Buildings 273:112367

253. Zölch T, Henze L, Keilholz P, Pauleit S (2017) Regulating urban surface runoff through nature-based solutions–An assessment at the micro-scale. Environ Res 157:135–144
254. Zölch T, Maderspacher J, Wamsler C, Pauleit S (2016) Using green infrastructure for urban climate-proofing: An evaluation of heat mitigation measures at the micro-scale. Urban Forestry & Urban Greening 20:305–316
255. Zuniga-Teran AA, Staddon C, de Vito L, Gerlak AK, Ward S, Schoeman Y, Booth G (2020) Challenges of mainstreaming green infrastructure in built environment professions. J Environ Planning Manage 63(4):710–732

The Key to Yazd's Resilience: Badgirs and Qanats, Innovative Secrets for Coping with the Arid Climate

Saeedeh Moayedfar and Seyedeh Saeedeh Hoseinyzadeh

1 Introduction

The accelerating impacts of climate change—manifesting as rising temperatures, prolonged droughts, and acute water scarcity—have posed existential threats to urban centers, particularly within arid and semi-arid regions [1]. Conventional architectural and urban planning models, which rely heavily on energy-intensive mechanical systems, are increasingly evaluated as inefficient, vulnerable, and unsustainable in these shifting conditions [2]. Conversely, Traditional Ecological Knowledge (TEK) and vernacular building practices have emerged as time-tested solutions for climate adaptation and robust "energy resilience"paradigms. This body of knowledge provides a vital repository for designing low-energy, resilient urban futures [3–5].

In the discourse of architecture and urbanism, energy resilience refers to the capacity of built environments to maintain essential functions and human comfort in the face of climatic and energy shocks or stresses, with minimal reliance on external energy inputs [6]. This concept transcends mere energy efficiency, encompassing systemic pathways for long-term absorption, adaptation, and transformation [7, 8]. Within this context, indigenous infrastructures are increasingly interpreted as key providers of urban ecosystem services, such as passive thermal regulation and sustainable water management [9].

The historic city of Yazd, Iran—a UNESCO World Heritage site—serves as an outstanding exemplar of such indigenous-based energy resilience. For centuries, its inhabitants have endured a harsh desert environment characterized by a hot desert climate (Köppen classification BWh), extreme diurnal temperature fluctuations, negligible precipitation, and low humidity [10]. The city's climatic robustness

S. Moayedfar (✉)
Department of Geography, Faculty of Humanities, Meybod University, Meybod, Iran
e-mail: moayedfar@meybod.ac.ir

S. S. Hoseinyzadeh
Department of Art and Architecture, Payame Noor University, Tehran, Iran

A. Cheshmehzangi et al. (eds.), *Design Innovation for Climate-Resilient Cities and Built Environments*, Urban Sustainability, https://doi.org/10.1007/978-981-92-2297-1_3

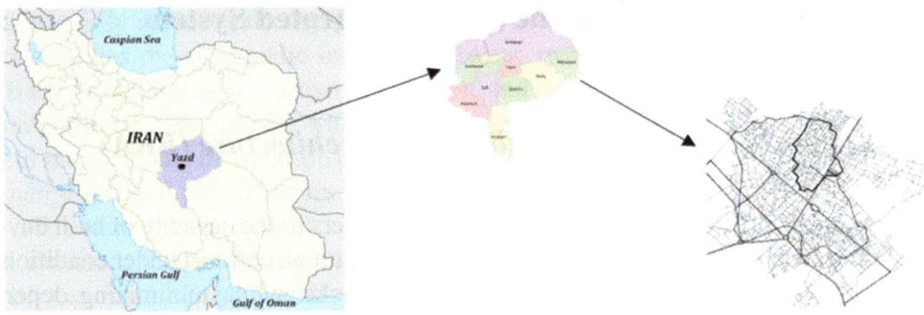

Fig. 1 Map of Yazd, located in central Iran

been transformed into a "living laboratory" for indigenous energy resilience and sustainable urbanism.

2.3 Case Study: The Badgir–Qanat Synergistic System

The integrated windcatcher–qanat system represents a tangible and sophisticated example of an energy-resilient SETS, intricately intertwining technical, ecological, and social dimensions. Within this framework, the technological-ecological synergistic function is driven by the continuous flow and evaporation of qanat water, which serves as a primary source for passive cooling and humidification [24]. This mechanism forms the conceptual basis for the "temperature and humidity regulation" criterion in the present analysis. By transporting cool water from distant aquifers, the qanat fulfills the dual critical functions of providing a sustainable water supply and creating a subsurface passive cooling source (Fig. 2).

Complementing this, the windcatcher (Badgir), through a sophisticated mechanism that operates effectively with or without ambient wind (Fig. 3), channels the

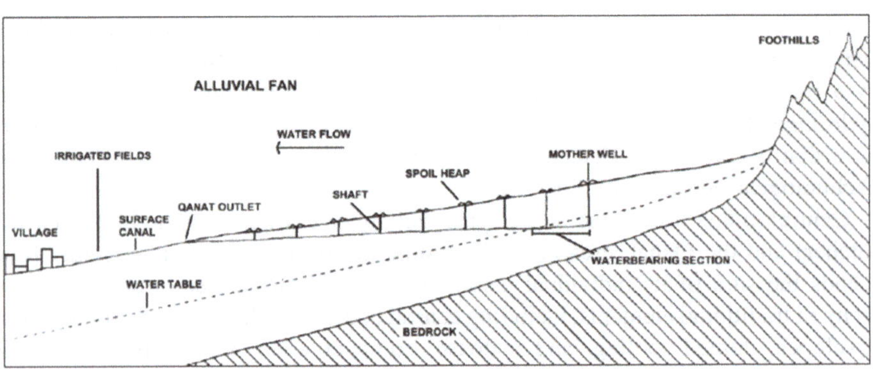

Fig. 2 Schematic section of qanat vertical shafts and underground conduit. *Source* [25]

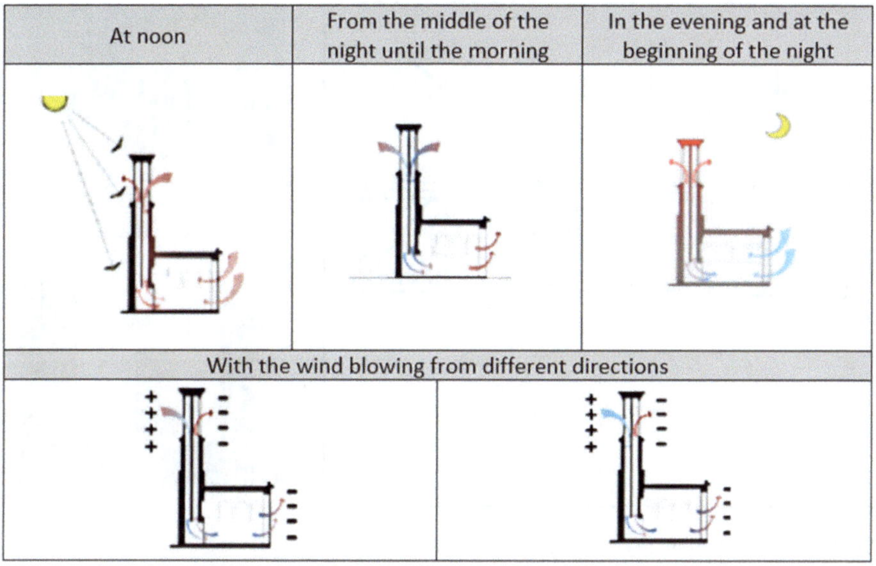

Fig. 3 Function of the Badgir in the absence of wind

cool and humid air from these subsurface spaces—such as the Qanat, Sardab (cellar), or Ab-anbar (cistern)—directly into the building interior [11].

This synergy facilitates high-efficiency natural evaporative cooling, which is capable of reducing indoor temperatures by as much as 20 °C relative to the external environment [26]. Furthermore, the diverse morphologies and typologies of windcatchers—ranging from unidirectional to four-sided and eight-sided configurations (Fig. 4)—reflect adaptive responses to prevailing wind directions and site-specific environmental conditions [27].

Beyond its technical merits, the system encompasses a significant socio-economic dimension; it transcends mere technology to constitute a social structure centered on the communal management of water and a collective body of knowledge regarding its construction and maintenance. While direct access to the qanat via a private *payab* or *sardab* often functioned as an indicator of economic prosperity—since private access was typically reserved for aristocratic residences [28]—the windcatcher served as a low-cost, passive cooling solution integrated into nearly all households [29]. Consequently, this synergy transformed energy resilience into a pervasive and democratic value within the society. Examples of traditional buildings where these elements harmoniously coexist are further detailed in Table 1.

The functionality of this system is further enhanced and reinforced by an integrated network of complementary architectural elements. These include the Sardab, a subsurface space designed for cool habitation and food storage, the Payab, which provides critical staircase access to qanat water, and the Ab-Anbar, a public water reservoir that utilizes windcatchers for cooling and ventilation (Fig. 5).

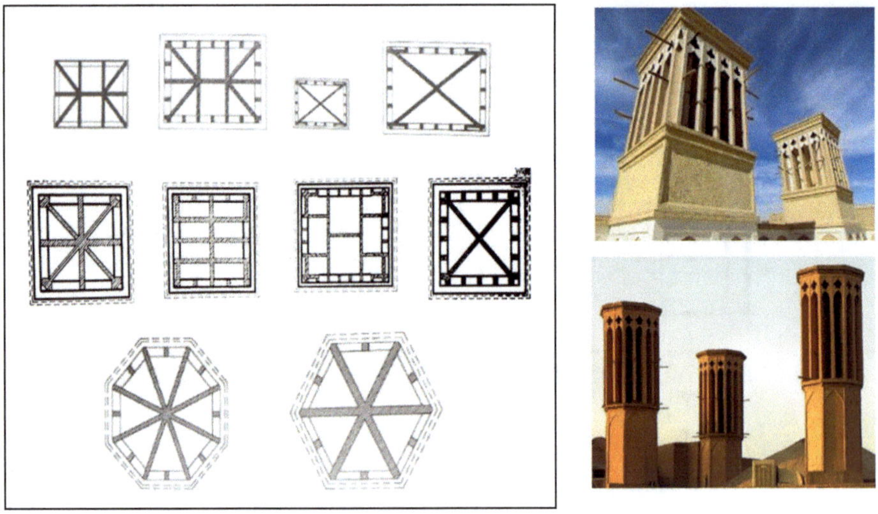

Fig. 4 Kinds of Badgirs in configurations

Table 1 Samples of traditional resilient buildings in Yazd

Buildings	Age	Kinds of Badgir	Building Performance recently	Repaired & usable	Place of Badgir in the plan of the house	Perspective	Place of Payab or Sardab in the basement
Tehrani house	170 years	Four sided	As a hotel-museum	☑	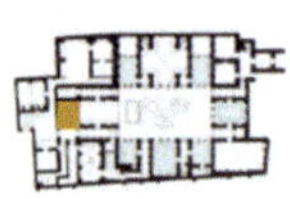	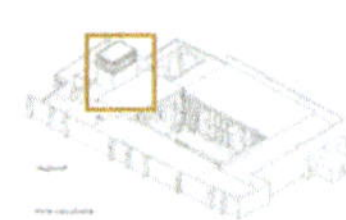	
Akhavan-Sigari house	100-150 years	One Four sided, the other two sided	As an Eco-lodge	☑	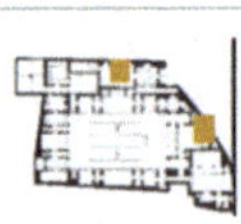	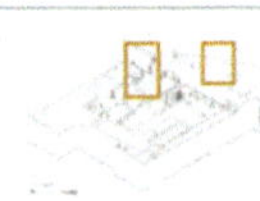	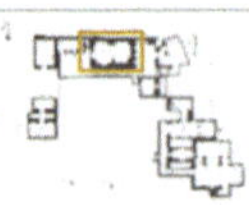
Dolat-Abad garden (Kushk)	250 years	Octahedral	As an Eco-lodge, traditional museum	☑	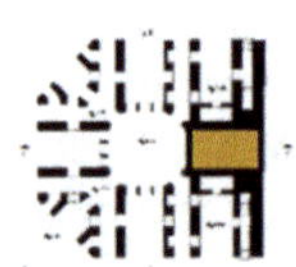	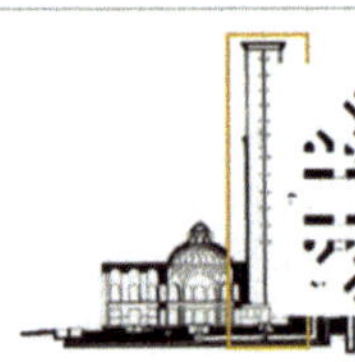	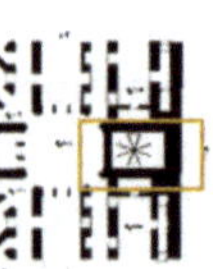
6 Badgirs Ab-Anbar	166 years	6 Octahedral Badgirs	As an Ab-Anbar	Abandoned	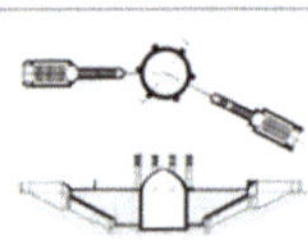	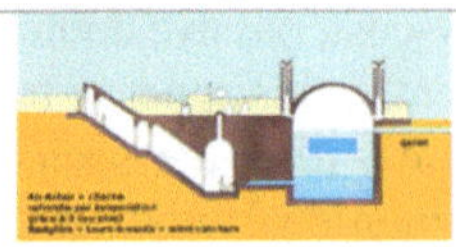	

Fig. 5 Payab and private pool to access the Qanat water in traditional houses in Yazd. Source: [29]

In conclusion, it can be asserted that in the traditional urban fabric of Yazd, the integrated operation of the windcatcher–qanat system has played a pivotal role in fostering energy resilience. This resilience is not merely the result of isolated components but rather the outcome of a cohesive, interconnected network where social, ecological, and technological dimensions overlap.

The specific roles of the key components of Yazd's SETS framework—and their respective contributions to energy resilience—are synthesized in Table 2.

Ultimately, a contemporary evaluation of this system necessitates a comprehensive framework capable of measuring such integration and synergy across its diverse dimensions. By adopting the SETS framework and employing the Analytic Hierarchy Process (AHP), the present study seeks to provide a systematic assessment that captures the complexity of these traditional adaptive strategies.

3 Research Methods

This study employs a mixed-methods and applied approach to prioritize the key performance criteria for the integrated windcatcher (Badgir) and Qanat system in Yazd. Given the multidimensional nature of resilience—encompassing

Table 2 Role of key elements in Yazd's SETS system for energy resilience

Element	Key function in energy resilience	Type of resilience capacity	Role in the integrated system
Badgir (Windcatcher)	Provides natural ventilation and evaporative cooling, reducing reliance on mechanical energy	Absorptive + Adaptive	Active distribution node: Channels air from the cooling source (often linked to the Qanat) into living spaces
Qanat (Subterranean aqueduct)	Supplies sustainable cool water, moderates soil temperature, and enhances ambient humidity	Absorptive + Anticipatory	Ecological Source & Engine: The foundational infrastructure enabling passive cooling and hydrological sustainability
Sardab/Payab (Cellar/ Underground water access)	Creates a cool subterranean space and provides direct access to Qanat water	Absorptive	Interface space: The physical and functional nexus connecting the Qanat (source) and the Badgir (distributor)
Ab-Anbar (Public cistern)	Stores and cools public water, creating a localized microclimate	Absorptive + Adaptive	System Buffer & Amplifier: Regulates water access and enhances the cooling effect through its integrated windcatchers
Central Courtyard & Urban Form	Provides shading, entraps humidity, and reduces the urban heat load	Adaptive + Anticipatory	Performance-enhancing matrix: The contextual framework that maximizes the efficiency of the core Badgir–Qanat synergy

social, ecological, and technological dimensions—a Multi-Criteria Decision-Making (MCDM) framework is deemed essential for a comprehensive evaluation. Accordingly, the Analytic Hierarchy Process (AHP) has been selected as the primary analytical model. AHP serves as a systematic method that integrates mathematical reasoning with expert judgment to organize and evaluate complex decision-making processes [30]. Its primary utility lies in its capability to decompose intricate problems into a manageable hierarchical structure, quantify expert judgments through pairwise comparisons to generate reliable priority weights for each criterion [31]. This framework is specifically utilized to handle the quantitative prioritization of the diverse factors influencing energy resilience in the study area.

Table 3 Effective criteria for climate resilience with emphasis on SETS factors

Factor	Explanation
Technological Efficiency & Adaptability	The performance and flexibility of the physical infrastructure in providing cooling, ventilation, and water
Ecological sustainability	The system's impact on and synergy with the local environment, including water conservation and microclimate regulation
Socio-cultural value	The role of the system in preserving identity, indigenous knowledge, and community cohesion
Economic feasibility	The cost-effectiveness and resource efficiency related to construction, maintenance, and operation
Structural Durability & Longevity	The inherent resilience of materials and design against climatic and temporal stressors

Source The review of the research literature, expert opinions, and emphasis on specific articles such as [9, 16].

3.1 Identification of Key Climate Resilience Criteria

The initial phase of the research was dedicated to identifying the fundamental criteria required to assess the climate resilience of the integrated windcatcher and qanat system. This process involved a systematic literature review focused on vernacular architecture, the hydro-thermal performance of qanats and windcatchers, and broader urban resilience frameworks, with a specific emphasis on the Social-Ecological-Technological Systems (SETS) perspective. To refine the findings from the literature, the review was enriched by preliminary insights gained through semi-structured interviews with three local heritage experts. As a result of this comprehensive synthesis, five primary evaluation criteria were established. These criteria were deliberately selected to capture the intertwined dimensions of the SETS framework, ensuring that the technical efficiency of the structures is evaluated alongside their ecological and social contributions (Table 3). Structurally, this model constitutes a two-level hierarchy: the overarching goal of assessing integrated energy resilience occupies the primary level, while the five identified criteria form the second level of the analytical framework.

3.2 Expert Panel Selection and Data Collection

The input data for the AHP were derived from a carefully selected panel of domain experts, ensuring that the prioritization process was grounded in high-level professional and academic knowledge. The target population comprised specialists and academics with proven expertise in the vernacular architecture of Yazd, sustainable water management, cultural heritage conservation, and urban climatology. To ensure the reliability of the judgments, a purposive sampling method was employed, specifically targeting individuals with a minimum of 10 years of professional or research experience in their respective fields. Of the 26 potential participants initially invited via email, 22 experts agreed to participate in the study.

This final expert panel ($n = 22$) represented a balanced interdisciplinary group, consisting of seven vernacular architects and architectural historians, five cultural heritage conservation specialists, six urban planners and climatologists, and four hydraulic engineers and qanat specialists. In terms of demographic and professional composition, the panel included 17 male and 5 female experts, with ages ranging from 30 to 50 years. Regarding their academic credentials, three members held a PhD and five held a Master's degree, while the remaining fourteen participants possessed Bachelor's degrees complemented by extensive field-based expertise in traditional construction and qanat management. This diverse composition was essential to capture both the theoretical and practical dimensions of the SETS framework.

Data collection was conducted using a custom-designed AHP pairwise comparison questionnaire, distributed and completed electronically. Within this instrument, experts were required to evaluate every possible pair of the five criteria based on the standard Saaty 9-point scale. This scale ranges from 1, representing equal importance, to 9, signifying extreme importance, as detailed in Table 4. This systematic approach allowed for the conversion of subjective expert insights into quantifiable priority weights for the subsequent analysis.

In the construction of these matrices, the reciprocal condition was strictly observed as a mathematical necessity. This principle ensures that if the importance of criterion i relative to criterion j is assigned a value of $a_{ij} = k$, then the relative importance of j to i must automatically satisfy $a_{ij} = 1/k$. This fundamental property was maintained across all levels of the hierarchy to ensure the internal consistency of the model.

Table 4 Intensity of importance table

Degree of importance	Definition	Explanation
1	Equal importance	Both factors have the same importance
3	Moderate importance	According to experience and judgment, one factor is more important than the other
5	Strong importance	One factor is more important than the other
7	Very strong importance	One factor is strongly preferred at a higher level than the other
9	Extreme importance	One of the factors is very important at a very high rate
2,4,6,8	Intermediate values	These are the intermediate rates; they use when compromise is needed

Source [31]

3.3 Data Analysis Using AHP

The data analysis phase commenced with the aggregation of the completed pairwise comparison matrices from the 22 experts. To form a single, representative comparison matrix, the geometric mean method was employed, which is the standard approach for maintaining the reciprocal property of the matrices in group decision-making. Subsequently, the relative priority weights for each criterion were derived through a systematic two-step process: first, the aggregated pairwise comparison matrix was normalized, and second, the principal eigenvector of the matrix was computed to yield the final criteria weights.

To ensure the reliability of the findings, the consistency of both the individual expert judgments and the final aggregated matrix was rigorously evaluated by calculating the Consistency Ratio (CR). Adhering to the established thresholds in AHP methodology, a CR value of less than 0.10 was deemed acceptable, confirming that the pairwise comparisons possessed sufficient logical coherence and were free from significant contradictions. All computational analyses were performed using Expert Choice 11 and Microsoft Excel. The final output of this analytical phase is a set of normalized priority weights, summing to 1.00, which provides a quantitative basis for determining the relative importance of the five criteria in fostering energy resilience.

4 Results

Following the AHP methodology detailed in the previous section, the aggregated pairwise comparison matrices were processed to derive the priority weights for the SETS dimensions. The numerical findings, illustrating the relative importance of each criterion, are summarized in Table 5.

The aggregated pairwise comparison matrix, representing the geometric mean of the judgments provided by the 22 experts, is presented in Table 5. To ensure the validity of these comparative assessments, the Consistency Ratio (CR) was calculated. The analysis yielded a CR of 0.048, which is significantly below the permissible threshold of 0.10 established by Saaty [31]. This finding confirms the logical coherence and reliability of the collective expert judgments, indicating that the comparisons are free from significant contradictions.

Following the verification of consistency, the matrix was normalized to determine the relative weight of each factor. The resulting normalized matrix, which serves as the basis for calculating the final priority weights, is detailed in Table 6. As shown in these results, there is a clear hierarchical distinction between the criteria, with "Technological Efficiency & Adaptability" showing a dominant influence compared to other dimensions.

Table 5 Aggregated pairwise comparison matrix of criteria (geometric mean of 22 experts)

Criteria	Technological efficiency & adaptability	Ecological sustainability	Socio-cultural value	Structural durability & longevity	Economic feasibility
Technological efficiency & adaptability	1	2	2	3	4
Ecological sustainability	1/2	1	1	2	3
Socio-cultural value	1/2	1	1	2	3
Structural durability & longevity	1/3	1/2	1/2	1	2
Economic feasibility	1/4	1/3	1/3	1/2	1

Table 6 Normalized matrix of criteria

Criteria	Technological efficiency & adaptability	Ecological sustainability	Socio-cultural value	Structural durability & longevity	Economic feasibility
Technological efficiency & adaptability	0.387	0.414	0.414	0.353	0.308
Ecological sustainability	0.194	0.207	0.207	0.235	0.231
Socio-cultural value	0.194	0.207	0.207	0.235	0.231
Structural durability & longevity	0.129	0.103	0.103	0.118	0.154
Economic feasibility	0.097	0.069	0.069	0.059	0.077

4.1 Analysis and Interpretation of Priority Weights

Following the normalization process, the final priority weights for the five climate resilience criteria were calculated by averaging the rows of the normalized matrix, as detailed in Table 7. The results reveal that Technological Efficiency & Adaptability (TEA) is the most critical dimension, accounting for 37.5% of the total weight (0.375). This is followed by Ecological Sustainability (ES) and Socio-Cultural Value (SCV), which hold equal importance at 21.5% each (0.215). In contrast, Structural Durability & Longevity (SDL) and Economic Feasibility (EF) were assigned lower priorities, with weights of 12.1% and 7.4%, respectively.

Table 7 Priority weights of climate resilience criteria for the Badgir–Qanat system

Criteria	Calculation (Sum of Row/5)	Final weight (Priority)
TEA	(0.387 + 0.414 + 0.414 + 0.353 + 0.308)/5	0.375
ES	(0.194 + 0.207 + 0.207 + 0.235 + 0.231)/5	0.215
SCV	(0.194 + 0.207 + 0.207 + 0.235 + 0.231)/5	0.215
SDL	(0.129 + 0.103 + 0.103 + 0.118 + 0.154)/5	0.121
EF	(0.097 + 0.069 + 0.069 + 0.059 + 0.077)/5	0.074
Total		1.000

This hierarchical distribution indicates that from the perspective of the expert panel, the energy resilience of the Badgir–Qanat system is primarily driven by its functional adaptability and technological performance. Interestingly, the equal weighting of ecological and socio-cultural dimensions underscores the balanced nature of these systems within the SETS framework, where environmental preservation and communal heritage are viewed as equally vital to long-term resilience. The relatively lower weight of economic feasibility suggests that in the context of Yazd's traditional architecture, the inherent value of climate adaptation and cultural continuity outweighs immediate financial considerations.

5 Discussion

This study aimed to identify and quantify the operational principles of climate resilience within the Yazd windcatcher–qanat system by integrating the Social-Ecological-Technological Systems (SETS) framework with the Analytic Hierarchy Process (AHP) multi-criteria decision-making method. The adoption of this framework directly responds to recent academic critiques emphasizing the necessity of shifting from purely conceptual applications of SETS toward operational, quantitative, and context-specific assessments [16]. By focusing on indigenous infrastructure in an arid climate, the present research addresses a critical gap in empirical SETS studies, particularly within the context of Global South cities.

The results reveal that Technological Efficiency and Adaptability, possessing the highest weight, constitute the core of the system's resilience. This finding aligns with established literature that characterizes indigenous arid-region infrastructures as passive, low-energy, and environmentally attuned climate technologies [11, 12, 32]. Within the SETS framework, such infrastructures are interpreted as key components in producing urban ecosystem services, specifically thermal regulation and sustainable water management [9]. Quantifying this component demonstrates that climatic performance remains the fundamental source of the system's adaptability in the face of intensifying climate change.

At the intermediate level of the hierarchy, equal weighting of Ecological Sustainability and Socio-Cultural Value (both 0.215) provides one of the most significant theoretical insights of this research. This result empirically confirms the foundational logic of the SETS, showing that in Yazd environmental and socio-cultural dimensions operate not in isolation but as intertwined and mutually dependent components. Such a finding resonates with studies emphasizing the role of indigenous knowledge in bridging ecology, technology, and meaning [3], while distancing the analysis from the common dichotomy between environmental conservation and cultural values in traditional heritage literature [33, 34].

Furthermore, at the lower levels of the hierarchy, the precedence of Structural Durability over Economic Feasibility (0.121 vs. 0.074) indicates that the resilience logic of this system is predicated on long-term investment and vulnerability reduction rather than short-term financial returns. This outcome aligns with critiques of economics-driven heritage conservation [18] and is consistent with literature on governing sustainability transformations, which prioritizes institutional and structural capacities for resilience [19]. Within this framework, economic factors are viewed as supporting conditions for continued functionality rather than the primary driver of decision-making.

From a methodological perspective, utilizing AHP to quantify multidimensional values enabled a systematic comparison among heterogeneous resilience components. This approach aligns with recent efforts to operationalize vulnerability and risk in SETS, particularly in urban water systems requiring integrated assessments [35]. Thus, the proposed model is applicable not only for analyzing indigenous heritage but also as a transferable tool for other climatic infrastructures in arid regions. Ultimately, this discussion reinterprets the Yazd windcatcher–qanat system as a living climatic infrastructure—one in which technological performance, ecological logic, and social acceptance operate in an integrated manner. This interpretation reflects contemporary urban planning approaches that emphasize the holistic integration of SETS [36], providing a robust foundation for the study's concluding remarks.

6 Conclusion

This study has demonstrated that the Yazd windcatcher–qanat system is far from a static historical relic; rather, it functions as a dynamic and resilient climatic infrastructure where technological performance, ecological harmony, and socio-cultural meaning are inextricably interwoven. By operationalizing the SETS framework through the AHP method, this research has successfully quantified and prioritized the multifaceted components of resilience, leading to several fundamental insights.

A primary conclusion is that technological efficiency and adaptability constitute the very core of the system's resilience. This finding underscores the critical importance of practical performance and technological responsiveness to climatic stresses, validating the functional essence of indigenous climate technologies. Furthermore, the empirical evidence reveals that ecological sustainability and socio-cultural value are inseparable and mutually reinforcing components. This reinforces the necessity of viewing urban heritage through the lens of integrated governance, where the environment, society, and technology are managed as a unified whole.

The prioritization of structural durability over immediate economic considerations further suggests that heritage conservation must be reframed as a long-term investment in urban resilience rather than a project driven by short-term financial returns. From a practical perspective, these findings argue that safeguarding such systems must extend beyond mere aesthetics or the preservation of physical fabric; instead, they should be integrated into the broader framework of climate-resilient urban infrastructure. Theoretically, this study provides a robust example of how conceptual SETS models can be transformed into operational tools within the context of indigenous architecture, effectively addressing documented gaps in the existing literature.

Ultimately, the Yazd windcatcher–qanat system offers a living model of urban resilience that can inform policymaking, urban design, and conservation planning in arid and semi-arid regions globally. It stands as a powerful testament to the fact that true heritage is not found in the past alone, but in systems that actively sustain and adapt to urban life. Its enduring vitality is proven through its continuous, synergistic integration with the ecological, technological, and social fabric of the city, providing a blueprint for sustainable urbanism in an era of global environmental change.

Acknowledgments We extend our sincere thanks to all individuals who contributed to the completion of this research, and I am grateful to the authors of the text for the opportunity they provided.

References

1. El Kenawy AM (2024) Hydroclimatic extremes in arid and semi-arid regions: status, challenges, and future outlook. In: Hydroclimatic extremes in the middle East and North Africa. Elsevier, pp 1–22
2. Munonye WC, Ajonye GO (2025) Energy-driven circular design in the built environment: rethinking architecture and infrastructure. Front Sustain Cities 7:1569362
3. Berkes F (2017) Sacred ecology. Routledge
4. Gall T, Allam Z (2022) Strategic foresight and futures thinking in urban development: reframing planning perspectives and decolonising urban futures. In: Future cities series: practical guidance for innovative, resilient and inclusive cities of the future
5. Reinfandt L. (2015). The City in the Islamic world
6. Elmqvist T, Andersson E, Frantzeskaki N, McPhearson T, Olsson P, Gaffney O, Takeuchi K, Folke C (2019) Sustainability and resilience for transformation in the urban century. Nat Sustain 2(4):267–273
7. Folke C (2016) Resilience (republished). Ecol Soc 21(4)
8. Graveline M-H, Germain D (2022) Disaster risk resilience: conceptual evolution, key issues, and opportunities. Int J Disaster Risk Sci 13(3):330–341
9. McPhearson T, Cook EM, Berbés-Blázquez M, Cheng C, Grimm NB, Andersson E, Barbosa O, Chandler DG, Chang H, Chester MV (2022) A social-ecological-technological systems framework for urban ecosystem services. One Earth 5(5):505–518

10. Moosavi R, Vakilinezhhad R (2021) Experimental evaluation of summer thermal comfort in various types of Sardab (cellar): underground space in Iran vernacular houses. J Heat Mass Transf Res 8(1):1–11
11. Bahadori MN (2018) Passive cooling systems in Iranian architecture. In: Renewable energy (p. Vol1_87-Vol1_101). Routledge
12. Fathy H (2014) Natural energy and vernacular architecture. In: Low: bold approaches, measured architecture.-(inside & beyond; 1), pp 28–32
13. Memarian G, Brown F (2004) The shared characteristics of Iranian and Arab courtyard houses. In: Courtyard housing. Taylor & Francis, pp 49–62
14. Naiemi AH, Yeganehfarzand S (2019) Nashtifan windmills in their environmental context, Khurasan, Iran. Vernac Archit 50(1):57–77
15. Meerow S, Newell JP, Stults M (2016) Defining urban resilience: a review. Landsc Urban Plan 147:38–49
16. Sharifi A (2023) Resilience of urban social-ecological-technological systems (SETS): a review. Sustain Cities Soc 99:104910
17. Manshour S, Lehmann S (2025) A systematic review of passive cooling strategies integrating traditional wisdom and modern innovations for sustainable development in arid urban environments. https://arxiv.org/abs/2507.09365
18. Pendlebury J (2013) Conservation values, the authorised heritage discourse and the conservation-planning assemblage. Int J Herit Stud 19(7):709–727
19. Krueger EH, Constantino SM, Centeno MA, Elmqvist T, Weber EU, Levin SA (2022) Governing sustainable transformations of urban social-ecological-technological systems. Npj Urban Sustain 2(1):10
20. Wilson A (2021) Passive survivability. In: Climate adaptation and resilience across scales, pp 141–152
21. Hollnagel E (2013) Resilience engineering in practice: a guidebook. Ashgate Publishing, Ltd
22. Pauleit S, Ambrose-Oji B, Andersson E, Anton B, Buijs A, Haase D, Elands B, Hansen R, Kowarik I, Kronenberg J (2019) Advancing urban green infrastructure in Europe: outcomes and reflections from the GREEN SURGE project. Urban For Urban Green 40:4–16
23. Gharipour M (2016) Contemporary urban landscapes of the Middle East. Routledge, London
24. Hamidian A, Ghorbani M, Abdolshahnejad M, & Abdolshahnejad A (2015). Retracted: Qanat, traditional eco-technology for irrigation and water management
25. Beaumont P (1971) Qanat systems in Iran. Hydrol Sci J 16(1):39–50
26. Abouei R (2006) Conservation of Badgirs and Qanats in Yazd, Central Iran, pp 6–8
27. Sangdeh PK, Nasrollahi N (2022) Windcatchers and their applications in contemporary architecture. Energy Built Environ 3(1):56–72
28. Moayedfar S, Hoseinyzadeh SS (2025) Characterization of healthy buildings in Yazd traditional houses. In: Designing healthy buildings and communities: shaping a climate-resilient future. Springer, pp 25–48
29. Soltani A (2017). From heritage to sustainable design: focus on traditional housing architecture in Iran: analyzes and recommendations for sustainable design in hot and arid region. Case study: traditional and contemporary housing architecture in Yazd, Iran
30. Şahin M, Yurdugül H (2018) A content analysis study on the use of analytic hierarchy process in educational studies. J Meas Eval Educ Psychol 9(4):376–392
31. Saaty TL (1980) The analytic hierarchy process: planning, priority setting, resource allocation. RWS Publications
32. Vellinga M (2015) Vernacular architecture and sustainability: two or three lessons. In: Vernacular architecture: towards a sustainable future, pp 3–8
33. Harrison R (2012) Heritage: critical approaches. Routledge
34. Smith L (2006) Uses of heritage. Routledge, London. https://doi.org/10.4324/9780203602263
35. Khromova S, Méndez GV, Eckelman M, Herreros-Cantis P, Langemeyer J (2025) A social-ecological-technological vulnerability approach for assessing urban hydrological risks. Ecol Indic 173:113334

36. Branny A, Møller MS, Korpilo S, McPhearson T, Gulsrud N, Olafsson AS, Raymond CM, Andersson E (2022) Smarter greener cities through a social-ecological-technological systems approach. Curr Opin Environ Sustain 55:101168

Geometry-Led Residential Design for Climate Resilience in Hot Semi-arid Cities: A Case Study of Lahore

Ayesha Mehmood Malik, Saima Gulzar, and Memoona Rashid

1 Introduction

Urban conditions also make climatic stress increasingly more pronounced as extreme temperatures combine with dense built environment and minimal natural cooling to aggravate heat exposure in residential neighbourhoods [1, 2]. Housing is the centre of climate vulnerability in hot semi-arid areas of South Asia where there is a prolonged summer season and cooling-based energy demand. Even in cities like Lahore, indoor overheating of residential buildings is a chronic occurrence as they, at the same time, increase the rising electricity demand and reinforce the increased reliance on mechanical cooling in times of extreme heat. The current housing development trends in most Pakistani cities are based on the need to cover the largest possible area of the plot and repetitive designs, frequently without much attention to climatic responsiveness. These spatial arrangements expand the amount of exposed roof and facade areas, limit the possibilities of ventilation, and limit shaded transitional areas, enhancing the effect of indoor heat build-up. The decisive role of building-climate interaction is represented by architectural geometry. The direct effect of solar gain, massing, depth of building, integration of the courtyard, and articulation of the facade are directly affected by orientation, massing, building depth, courtyard integration, and facade articulation [3, 4]. Contrary to technology-dependent approaches, geometry-based strategies are passive and self-sustaining in nature and have long-lasting performance advantages on a performance basis with minimum levels of operational energy. Nevertheless, although the concept of climate-responsive design is well established, its contribution to thermal and energy resilience, in the sense of the ability of housing

A. M. Malik (✉) · M. Rashid
School of Architecture, Faculty of Arts and Architecture, University of Lahore, Lahore, Pakistan
e-mail: Ayesha.mehmood@arch.uol.edu.pk

S. Gulzar
School of Architecture and Planning, University of Management and Technology, Lahore, Pakistan

A. Cheshmehzangi et al. (eds.), *Design Innovation for Climate-Resilient Cities and Built Environments*, Urban Sustainability, https://doi.org/10.1007/978-981-92-2297-1_4

to be thermally functional in extreme heat and situations of energy stress, has in most cases remained implicit in research and practice [5]. This chapter thus uses the Geometric modelling to determine the effectiveness of residential design strategies using geometry as the determinant to improve thermal and energy resilience during hot semi-arid climates as a case of Lahore. Parametric modelling and environmental simulation are used to assess alternative spatial arrangements in comparison with a set base and map out to scalable architectural principles and policy-relevant recommendations on climate-resilient housing.

1.1 Climate Change and Urban Heat Stress in South Asia

South Asia is one of the world regions the most vulnerable to climate change, where the average temperature increases alongside the frequency and duration of significant heat events [6]. Urban heat island, high density of the built environment, a large amount of impermeable surfaces, and a lack of vegetation exacerbate the urban heat island phenomenon in cities and limit the capacity of buildings to cool at night, causing chronic thermal stress in residential settings [1, 2]. These conditions, in hot semi-arid cities, where climatic cooling potential is already low, reduce to continuous overheating of indoors and increasing dependence on mechanical cooling. Peak cooling demand in heatwave phases frequently coincides with grid pressure, exposing households to unsafe indoor temperatures in case of power outages. This convergence of climatic and infrastructural risk, in terms of resilience, demonstrates the need to minimise cooling dependency by passive design instead of depending on the growing supply of energy.

Figure 1 traces the demand of urban cooling in the context of overall regional energy and emissions trends and shows how development patterns that can be characterised as heat-intensive add to the increasing carbon footprint and systemic susceptibility. Though it is not an immediate measure of the severity of urban heat islands, the figure supports the argument that low-energy housing formats need to be adopted in the form of form-based housing strategies as a component of climate-adaptive urban transition. Exposure to heat in homes is also a social equity and health issue. Households experiencing low access to passive cooling capabilities or stable energy capacity are at higher risk of indoor overheating and are prone to heat-related disease and decreased thermal safety in the event of extreme circumstances [8, 9]. It is possible to point out the importance of spatial strategies that can moderate heat gain without any mechanical system. The South Asia climate adaptation frameworks are aware of the magnitude of investment needed to mitigate the risks posed by climatic conditions and a significant portion of funds is channelled towards infrastructure and the energy systems. Table 1 presents estimated costs of climate financing in selected countries that lean towards large-scale sectoral interventions. Although vital, such investments are not sufficient to tackle residential heat exposure; therefore, the use of complementary, low-cost building-scale interventions that may be applied to extensive housing stocks is supportive of accessing the same.

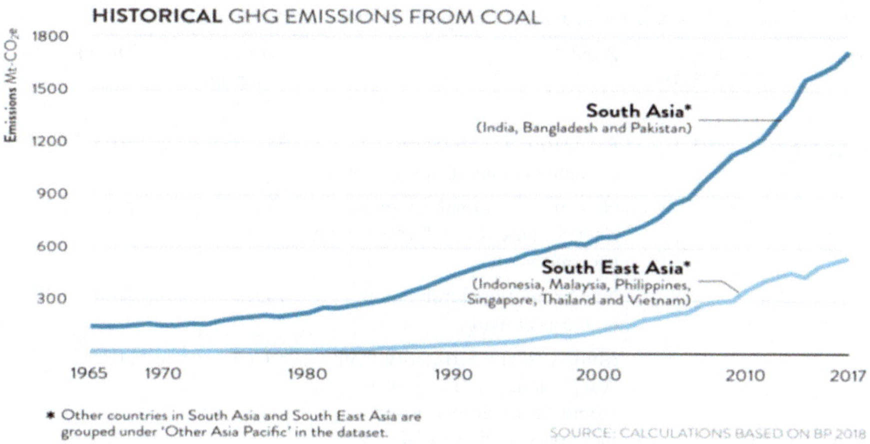

Fig. 1 Regional greenhouse gas emissions linked to energy-intensive urban development and cooling demand in South and Southeast Asia. *Source* [7]

In this regional setting, the form of residential buildings has become an important interface of climate-exposure, energy-vulnerability. Geometry-based design strategies can minimise the base cooling requirements and make housing more resilient to climatic and infrastructural stresses by minimising solar heat gains and maximising the potential of ventilation to make buildings thermally operative during times of climatic and infrastructural challenge. This makes architectural geometry a key element of climate-resilient urban development in the hot semi-arid environment.

1.2 *Urbanisation Patterns and Residential Heat Exposure in Lahore*

Rapid urbanisation in Lahore has transformed the morphology of residences in such a manner that the passive cooling potential is increasingly being constrained. Contemporary housing developments focus on maximum plot coverage, reduction of setbacks and compact street widths, which leads to dense building configurations with limited ventilation corridors and little shaded outdoor space. These spatial patterns diminish the possibilities for air movements, enhance surface heat storage, and also amplify indoor heat accumulation during long summer periods. Over time, residential plots have suffered progressive loss of open space and vegetation along with increases in building height and density (i.e., closer building spacing). This transformation has lessened the effect of traditional microclimatic buffers such as courtyards, rear yards, and shaded transitional zones that historically had a moderating effect on the indoor temperature. As a result, housing forms are becoming

Table 1 Estimated climate finance needs for selected South Asian countries

Country	Total (USD)	Breakdown	Cost of Component (USD)	Time Frame
Bangladesh	303.7	BPD2100: Adaptive water management and infrastructure	38	2018–2030
		NDC unconditional mitigation: Energy, agriculture, forestry, land use, and waste	32.3	2021–2030
		NDC conditional mitigation (same as above)	143.7	2021–2030
		Mujib Climate Prosperity Plan: Adaptation, just transition of labour, technology transfer, increased public revenue, renewable energy, energy efficiency, power, and transport resilience	89.7	2022–2030
Nepal	243.5	Adaptation	47.4	2021–2050
		Mitigation	196.1	2021–2050
Pakistan	348.4	Deep decarbonisation:—Low-carbon power supply ($84.7b)—Transport ($57b)—Phase out coal and replacing with solar ($31b)—Energy Efficiency (industry) ($9.8b)—Energy Efficiency (buildings) ($5.5b)—Clean cooking ($1.3b)—EE irrigation ($0.3b)	189.6	2023–2030
		Adaptation:—Disaster preparedness and response ($85.7b)—Universal water and sanitation ($55.2b)—Modernisation of irrigation systems ($4b)	144.9	2023–2030
		Crosscutting:—Wastewater management ($7.5b)—Municipal solid waste management ($6.4b)	13.9	2023–2030

Source [7]

less able to control thermal conditions by spatial means and increasingly reliant on mechanical cooling.

Table 2 summarises the shift in the residential developments in Lahore that have progressively moved towards increasing the plot coverage and standardised front lawn configurations at the expense of shaded internal open spaces and ventilation

Table 2 Change in land cover and built-up density in Lahore House from 2000 to 2020

Period	Representative Housing Scheme	Plot Diagram	Open Space Pattern	Built-Up Coverage	Cooling Implication
2000s	DHA EME (2001)	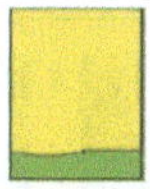	Front + rear lawns	65–70%	Good cross-ventilation, shaded buffer zones
2010s	Valencia (2016)		Front lawn dominant	64–70%	Limited airflow, increased façade exposure
2020s	Bahria Town (2020)		Front lawn only	62–65%	High cooling dependency, minimal passive control

Source Author

corridors. As can be seen in Table 2, previous housing projects normally incorporated front and back open areas, which served to promote air circulation and thermal insulation, but recent works have been dependent on almost costly front lawns. The given change restricts the possibility of cross ventilation, and exposure of facades to solar radiation, which in turn exacerbates the effect of heating indoors and reliance on mechanical ventilation.

2 Climatic and Microclimatic Conditions Influencing Residential Thermal Performance in Lahore

Lahore lies in the hot semi-arid climatic zone (BSh) characterised by extended periods of high temperatures, high solar radiation and low cooling (nocturnal) of night temperature in summer months [10]. Daytime temperatures rise often above 40 °C during the period of May to July, and the high night time temperature limits the release of the stored heat from the envelopes of the buildings. These conditions lead to ongoing overheating of buildings inside and make cooling the prime factor determining the residential energy demand [6]. The seasonal climatic changes are not of much use as far as thermal relief is concerned. The monsoon periods though temporarily lower ambient air temperatures, due to increased humidity often maintain indoor discomfort and negate the effectiveness of natural ventilation [11]. As a result, they have residential buildings that are left with large cooling loads throughout the year, further promoting reliance on mechanical air-conditioning systems and exposing them to greater vulnerability during times of peak electricity demand.

Figure 2 summarises the climatic drivers of Residential Heat Exposure in Lahore that are typical of this region for which the long-term meteorological normals for temperature, solar radiation, and wind speed were combined. These circumstances suppress the convective dissipation of heat through building surfaces and

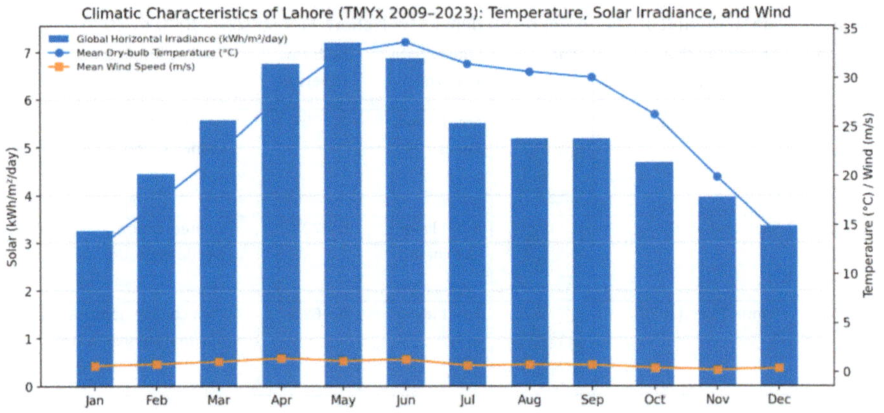

Fig. 2 Climatic characteristics of Lahore (WeatherSpark, n.d.)

amplify the radiative gain through the building, and it is consequently imperative to implement spatial strategies that minimise the building envelope, and provide better shading and ventilation routes [12]. In addition to the regional climate drivers, it has some local microclimatic conditions that have been added to the effects on residential thermal performance. Extensive roof exposure, high percentage of impervious ground surfaces and decreased vegetative cover promotes heat storage in the urban fabric and night-time cooling delay leading to high ambient and indoor temperatures [2]. This accumulated heating would result in a reduction in the availability of nocturnal cooling and increase the duration of indoor thermal discomfort.

Figure 3 shows the effects of high surface exposure and low shading on neighbourhood scale, which leads to higher levels of radiative heat absorption and lower levels of airflow potential. These microclimatic conditions enhance thermal loads on residential envelopes and have a stronger role on the indoor environment performance due to the influence of building geometry. The interplay between climatic exposure and surface characteristics of the urban area leads to a later release of the heat from building materials and the surrounding ground surfaces, especially in evening and night-time hours. This process extends indoor overheating beyond the onset of a fall in outdoor temperature letting people rely on constant mechanical cooling [1].

In addition, the potential contribution of cumulative storage of heat in a dense urban structure to long-term thermal stress indoors. In terms of resilience, these conditions negatively affect the ability of housing to survive heatwaves and energy outages, which supports the necessity of spatial planning that restrains the increase in solar energy and encourages the passive cooling of heat, [5]. The climatic and microclimatic conditions described above were used in the simulation model to inform the parameters used in this study for the performance comparison to be realistic, i.e., reflecting real exposure scenarios for residential buildings in Lahore. This section presents the climatic rationale of the study by defining cooling-dominated boundary conditions and ongoing heat stress as the major environmental challenge and defining

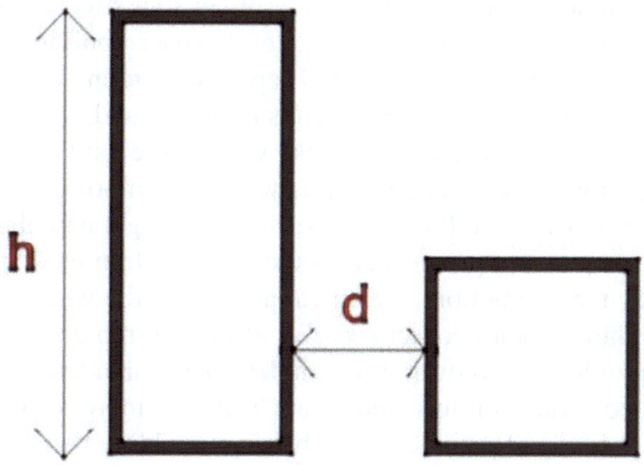

Fig. 3 Typical dense residential housing morphology in Lahore with limited open space and high surface exposure. *Source* [13]

the geometry and spatial layout of buildings as the primary determinants of thermal and energy resilience, which is elaborated on the theoretical framework below.

3 Theoretical Framework

This chapter preconditions the residential geometry as one of the strategies of adaptation to climate. It draws on climate responsive design, building physics and urban microclimatology to demonstrate the impact that the form and configuration of homes can have on thermal and energy performance at the scale of the single building and right up to neighbourhoods. Since ancient times, architecture in hot conditions has been based on spatial strategies such as compact blocks, shaded courts, tight streets, and well-placed openings in facades to regulate the amount of sun that enters and to encourage natural ventilation [3, 14]. These typologies engineer microclimatic gradients that limit heat build-up and increase comfort without the use of mechanical systems, making them inherently resilient to extremes of weather and infrastructural stress. However, in modern residential developments these traditional hints are usually discarded in favour of plot-based high-rise, high-floor-area developments. Geometric discordance between buildings, open space, and streets results and dilutes passive cooling potential and requires a heavier dependence on active cooling. This trend points to the much-needed re-establishment of geometry as the main driver of residential environmental performance in the larger climate resilience discourse.

3.1 Spatial Configuration as a Design Variable

The city morphology is the mute controller of the microclimate of a city, balancing the light, wind and heat in each block. Planners create the canyons where any structures trap heat or funnel breezes by organising streets, building height and distance. To illustrate, a series of tall and closely spaced buildings may create a narrow alley that reflects the rays of the sun, keeping the ground cool and may create a chimney-like effect in which hot air ascends to the ground. On the other hand, a more open grid with more staggered setbacks will enable daylight to enter deeper into the streets, promoting natural cooling as well as promoting cross ventilation, which dissipates heat. The overall outcome such as making a neighbourhood look like a dry canyon or a pleasant promenade would not be determined by the architectural variables but by the interaction of such variables with the winds and the direction of the sun. By taking a more conscious level of design into account when orienting streets, at the plot level, and overall, urban designers can attempt to reduce heat islands, promote pedestrian comfort, and create healthier, more resilient communities.

In Fig. 4 the interaction between building height, spacing and orientation can be seen to influence both the amount of solar infiltration and the amount of airflow across a neighbourhood. The diagram makes clear that urban form controls microclimatic conditions which determine levels of residential comfort and cooling demands. The street-level temperatures of neighbourhoods can be brought down by relatively minor geometric adjustments to individual plots, which can be a slight change in height or setback, or the facing direction of the façade, and which reduce cooling loads at the city scale overall.

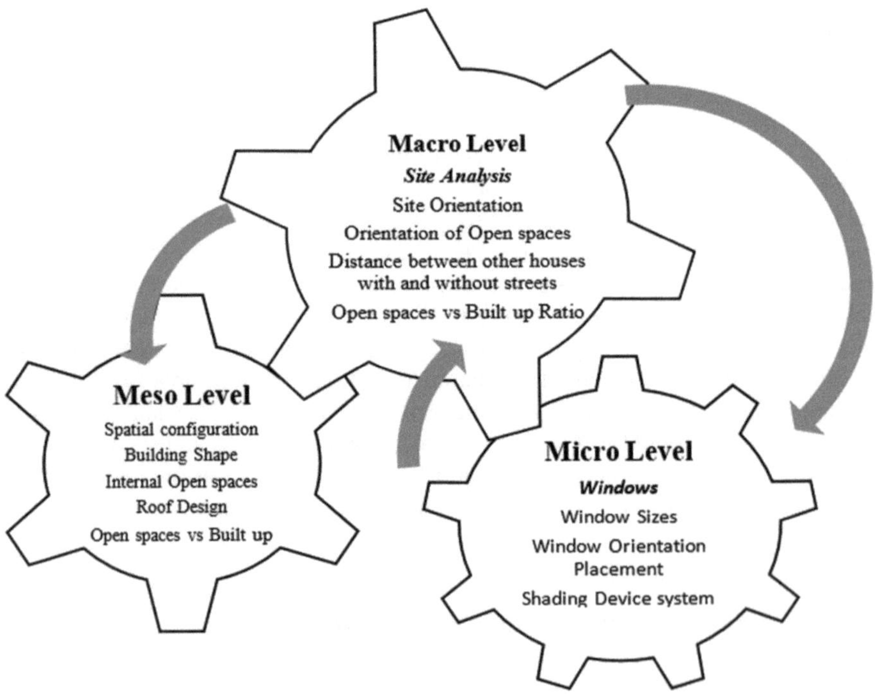

Fig. 4 Conceptual framework linking spatial configuration to thermal performance. *Source* Author

3.2 Building Geometry and Passive Thermal Regulation

When the shape and orientation of a building are taken into consideration, they become the main regulators of how much sun the building absorbs, how heat moves in and out, and how well the building can cool itself naturally. The orientation determines which facades receive the low-angle light of the sunrise and sunset, which affects the glare and thermal gain. Surface area-to-volume ratio tells us how much envelope is exposed to allow heat to move in or out; a high ratio means a lot of conductive and radiative heat transfer, which increases cooling loads. The Courtyards serve as interior shade islands, which reduce the microclimate: they promote the effect of stack-related ventilation, keep walls surrounding them cooler, and allow heat to escape at night. Shading devices like overhangs, louvres, shade screens intercept the direct solar rays before they fall on the envelope, evening out the indoor temperatures, reducing heat gains. All of these geometric strategies make a building that is cooler, consumes less energy, and can be more comfortable to be in Mashaly and Fahmy.

Figure 5(a) and (b) present a series of residential designs by their orientation to the street, mass organisation, the opening to the courtyards, and the use of shading. These types of designs are the building blocks behind the parametric studies that will follow, and they demonstrate that even small changes in the shape or orientation of a building can have a powerful effect on the amount of sun that the building obtains and the level of air circulation within the building. In the resilience perspective, these geometry decisions are made to keep the indoor environments comfortable in case of prolonged heatwaves, reducing the dependence of air-conditioning. When these measures are implemented on a neighbourhood-wide scale, they are able to reduce the total cooling load in the neighbourhood and improve the energy balance of the city.

3.3 Geometry-Led Design and Thermal-Energy Resilience

Geometry-based design is the bridge between building structure and the climate resilience of architectural adaptive and responsive thermoregulation instead of the post-hoc, energy-intensive retrofits. Parametric modelling is the basis of this approach, where designers can explore the relationship between orientation, massing and spatial configuration that affects thermal and energy performance iteratively and from the earliest stages of designing—when the geometry of the building establishes the path towards its long-term behaviour [15]. In a resiliency framework, residential performance should be evaluated not only in terms of efficiency in terms of normal conditions but also in terms of the capacity of a dwelling to limit indoor overheating during heatwaves and reduce reliance on mechanical cooling when the power grid is overloaded [5]. Geometry-based plans reduce the cooling needs at the

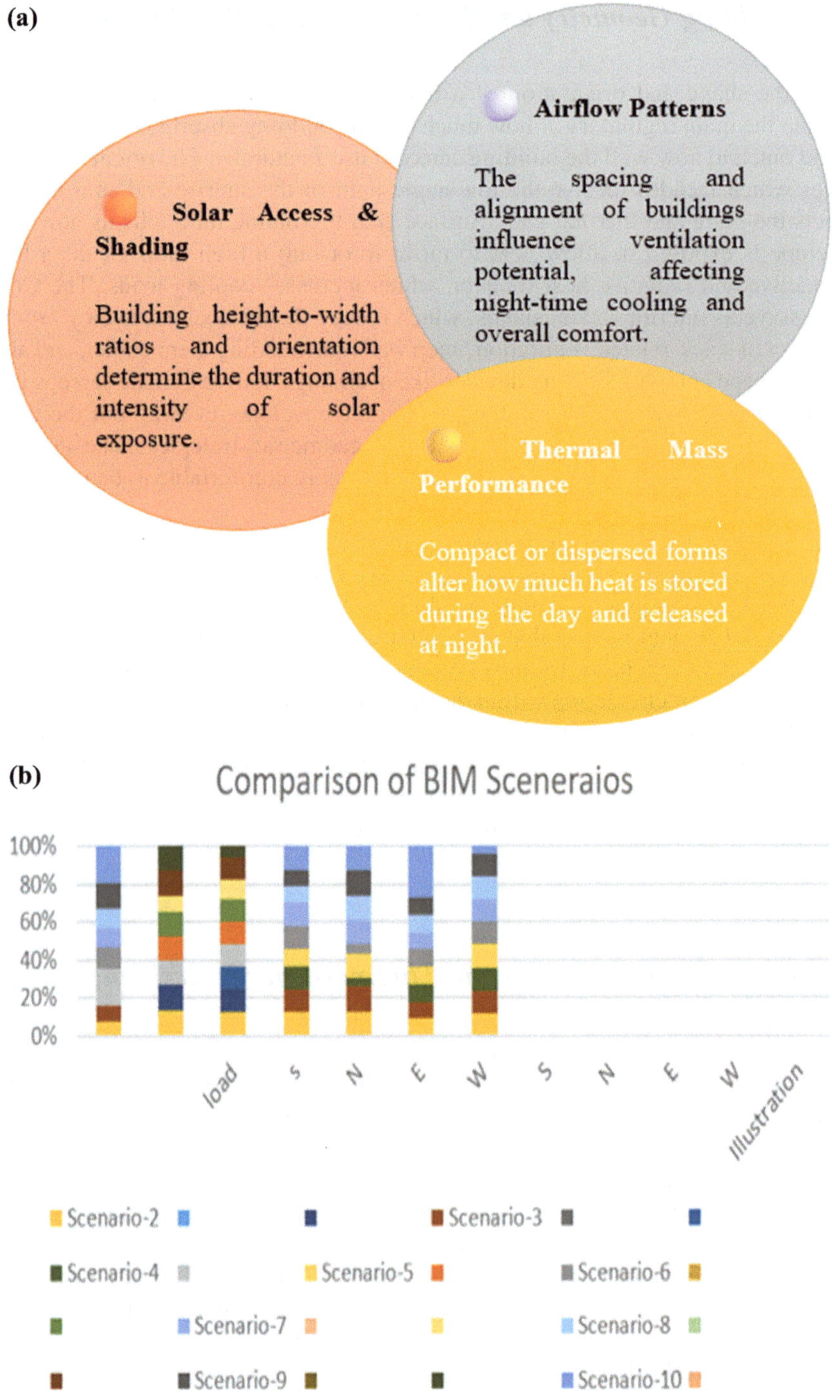

Fig. 5 (**a**, **b**) Parametric variations of spatial configuration in simulation models. *Source* Authors

base level, prolong the period of time that the interiors of structures can be comfortable without air-conditioning, and reduce peak power consumption by manipulating solar gains, airflow, and promoting nocturnal heat loss. When applied through a form base intervention, such interventions reduce anthropogenic heat release and help provide urban energy stability—as part of a wider heat mismatch mitigation strategy at neighbourhood and city scales.

4 Methodology

This paper is a parametric and simulation-based comparison, which is conducted to examine the relationship between shape and layout of residential buildings and cooling energy consumption, in a hot semi-arid climate. We do not wish to make forecasts on the precise consumption values, but instead we would like to rank alternative spatial arrangements with the same, controlled boundary conditions. This allows us to use geometry as a passive resilience mechanism that will reduce the demand for cooling, while keeping indoor temperatures more comfortable. Lahore was selected as the test site because of its long and hot summer season, high residential energy consumption, and the high density of residential buildings. We fed the simulations with typical meteorological year (TMY) data which well represents the weather of Lahore, by using the same input for all the scenarios so that the result can be compared with each other directly. First, we developed a baseline model of a typical single family dwelling that occurs in planned neighbourhoods of the city of Lahore. The base case includes local construction: reinforced concrete framing with masonry infill, typical roof assemblies, and typical envelope materials. Glazing ratios, floor to floor heights, and internal zoning were fixed in all runs ensuring that any performance differences are due only to changes in building geometry.

Figure 6 illustrates the parametric modelling process starting with an initial (baseline) residential configuration and moving through the process of making systematic changes to important geometric variables. Most influential parameters studied were orientation, massing proportions, surface area vs. volume ratio, placement of the courtyard, shading depth of the facade, and fenestration distribution—factors that are typically determined from the earliest stages of the design process and consequently are critical to low energy optimisation. Each parametric scenario included some controlled change to one or more of these attributes while keeping the other variables constant. Environmental performance was evaluated in Autodesk Revit using Insight 360, which combines simulation of dynamic energy modelling to calculate the cooling demand based on the geometry of the building, envelope properties, and operational schedules. We used the standard occupancy profiles, internal heat gains and HVAC set-points of the Insight 360 residential templates throughout all scenarios. Although the platform allows for quick comparative evaluation, the treatment of airflow and representation of thermal mass is simplified, thus the resulting performance is viewed as a relative indicator of performance rather than absolute energy predictions. To enable easy comparison, the simulation outcomes were normalised to the baseline model and reported as percentage change in cooling energy demand. This comparative-performance model helps to make meaningful cross-design comparisons and draw generalised design advice, as opposed to case-specific optimisations. While the properties of the envelope were held constant to a great extent, a few instances of wall construction were also tested to get an understanding of secondary effects against geometry.

Table 3: The following lists briefly the geometric variables that we have discussed and the reasoning behind our choice. Through mapping the design choices to the performance indicators which we observed directly, the table explains what it is quite clearly, why a specific shape or layout causes a specific result. The parameters chosen to reflect the type of plots and building footprints developers are faced with in Lahore-like, thus the lessons are directly applicable when applied to on-site design, as well as the housing regulations that the regulators develop. The close relation between the spatial design, empirical findings, and applicability to real-life situations reinforces the study to be relevant to the architects, planners, and even the policymakers.

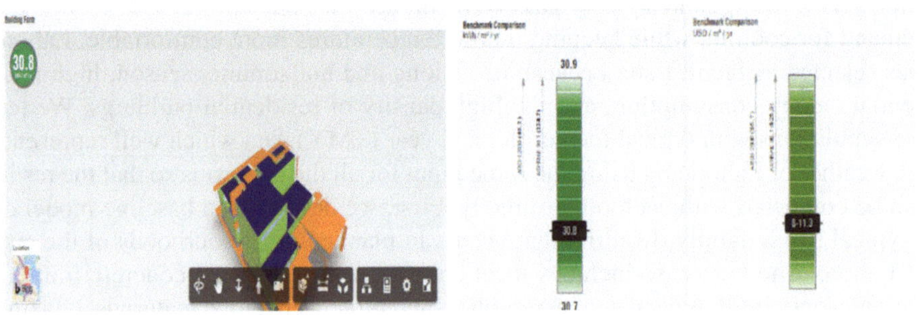

Fig. 6 Workflow of modelling and simulation process. *Source* Authors

Table 3 Geometrical parameters applied in simulation environment

Variable	Options Tested	Rationale
Orientation	North–south, east–west	Orientation affects solar gain and cross ventilation
Courtyard placement	Central, rear, lateral	Courtyard modifies microclimate, ventilation, shading
Built–open ratio	60%, 70%, 80% coverage	Determines heat gain and urban heat island effect
Shading devices	Overhangs, vertical fins, vegetation	Reduces solar heat gain on façades
Fenestration ratio	10%, 20%, 40% window to wall	Balances daylighting and cooling demand
Envelope materials	Brick cavity wall, RCC frame, insulated wall	Determines thermal conductivity and time lag

Source Authors

5 Findings

Simulations leave no crystal-clear account of how a house is formed and organised, being much more than a question of beauty: a major lever in reducing cooling loads and making interiors bearable to Lahore with its hot, semi-arid climate. With each design variation pitted against a well-defined baseline, we would be able to isolate the precise heat saving punch of each geometry strategy. Both of the most powerful moves are orientation and compactness. When a house is set in the east–west direction and the main facades are turned to the north and south, it is not merely keeping in order; it is also cutting off the sun, which is low and fiery in the mornings and late afternoons, striking east and west ways. The resultant decrease in direct solar gain is a 20–25% reduction in total cooling load, and maximum loads that peak during heatwaves are reduced significantly. The form of the building, which is made smaller, meaning reduced surface to volume ratio, further suppresses heat ingress. A smaller mass also implies that the sun has less surface area of the building envelope to impact, resulting in lower heat gain and reduced thermal inertia, which helps maintain more stable indoor temperatures. Courtyard Integration The Natural Ventilator. When this small size is combined with a balanced attic yard, good profits are increased. A central courtyard shades the core of the house, provides a channel for cross ventilation and allows hot air to escape at night. The cooling requirements were reduced by 22–28% over the baseline, given orientation and contextual massing, with the addition of levelling temperature changes during extended heatwaves, which is courtyard-based. Simply put, considerate geometry, primarily orientation, compactness, and interwoven court, is the source of both savings per year of energy and resistance to extreme heat, which proves that smart design is one of the best solutions to remain cool in the summers of Lahore.

Figure 7 demonstrates the effect of spatial layout on thermal performance. The designs built in courtyards remain cooler than high density block designs, despite the lack of mechanical cooling, due to their inherent passive survivability. This is critical in extreme heat spikes as overheating can be life-threatening to occupants. It is also revealed that the facade design has a difference: fixed horizontal overhangs and vertical fins are most effective on the building when they shield the sun hot sides, which shut out the blazing summer sun. At about 20% window-to-wall ratio, the daylight to temperature control is balanced, and a balance is achieved between excessively glazed and excessively minimalistic windows. Envelope materials alone provide small payoffs but combined with intelligent geometry they increase thermal lag and reduce peaks, which shows the importance of coordinate the materials selection with the spatial strategy. Building on these insights two residential prototypes were drafted to be put into real world application. The former, orientated east–west, compressed, and a central court mean the most of shading and natural draught. The second one is a small low-rise cluster that minimises exposure of envelopes and shadows facades and preserves functional layouts of small plots. To these two prototypes demonstrate that geometry-first designs may be scaled to a wide range of plot sizes and budgets, and provide large cooling-reduction advantages over traditional

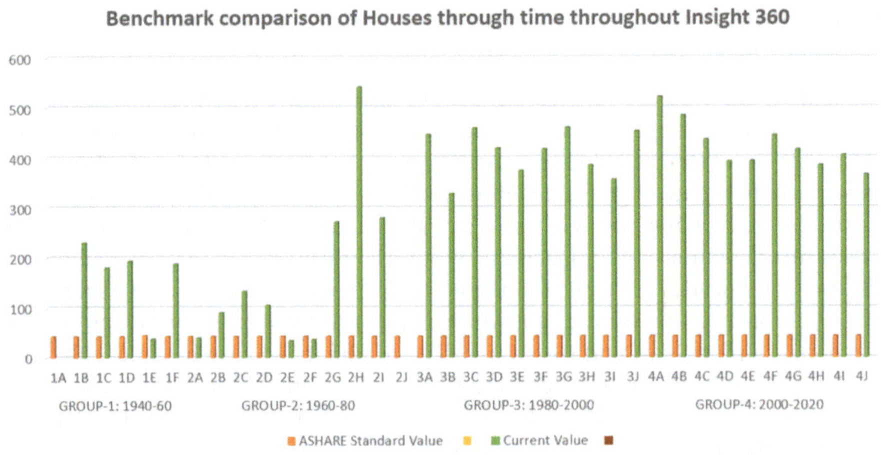

Fig. 7 Comparison of temperature profiles for different spatial configurations. *Source* Authors

designs. Concisely, the study establishes that a careful spatial planning by itself can bring about resilient and low-energy resilience in hot semi-arid cities.

6 Geometry-Led Design Principles for Climate-Resilient Housing

Based on the performance of trends determined in Sect. 5, this section will convey outcomes of the simulations into real architectural advice that can be applied to the design of a residential building in hot semi-arid climates. Instead of reproducing numerical outcomes, attention is paid to spatial strategies, which proved to be resilient repeatedly, in terms of heat gain reduction, usability of indoor temperatures, and a decreased reliance on mechanical cooling.

6.1 Core Spatial Principles

In all of the simulated configurations, there were four overlapping geometrical principles that were found to be the key to both thermal and energy resilience. To begin with, the control of orientation was essential: the residential plans were oriented into the east–west direction, the main facades were oriented north–south, which regularly minimised the effects of low-angle solar radiation and minimised the need to cool the building. Second, compactness in form with optimised surface-volume ratios were minimised to limit the envelope heat exchange and thermal storage, especially in use with shading of the facade. Third, courtyard integration brought about shady inner

Table 4 Geometry-led design guidelines and resilience implications for key housing stakeholders

Scale	Key Design Parameters (Case Study)	Performance Contribution
Macro (Plot)	Lahore; 418 m^2 plot; N–S orientation; 66% built/34% open; roof garden, trees, pergola, water bodies	Reduced solar exposure, shaded outdoor microclimate, lower surface temperatures
Meso (Form)	2 storeys; compact massing; GF 7 + FF 2 rooms; shaded terraces (NE/SE); exposed envelope ≈143 m^2; 9″ solid walls, perforated terrace walls	Lower envelope heat gain; improved ventilation and thermal lag
Micro (Envelope)	Strip latch windows; N–S openings; 20–24% window-to-wall ratio; horizontal shading with drop; internal doors 3′ × 7′ & 2.5′ × 7′	Controlled solar gains, balanced daylight, enhanced airflow

Source Authors

microclimates that improved ventilation and heat release in the night, which elevated thermal stability in interiors in prolonged heat blows. Fourth, moderated fenestration and orientation-responsive shading minimised direct solar gains and ensured functional daylight, and moderately high window-to-wall ratios worked better than highly glazed or minimally glazed facades. With these principles, it is possible to achieve the states of resilience in case of heatwave and energy stress without requiring active intervention and its implementation. Notably, the findings indicate that individual strategies are not enough but great benefits are achieved when orientation, massing, courtyards, and shading are combined at the initial phases of design.

Table 4 has listed the important geometric parameters and their respective resilience functions, which offer a brief design guide to architects and developers of houses. The principles are focused on form-based judgements, which might be made without using complicated technologies and expensive materials, which would help to maintain their scalability to various income levels and types of housing.

6.2 Prototype Translation of Design Strategies

To illustrate the applicability of architecture, two residential prototype setups were instantiated on the basis of the parametric scenarios that were the best performers. The initial prototype is made by taking the east–west orientation, compact massing, and centrally located courtyard in order to achieve maximum mutual shading and cross ventilation. This structure aids in maintaining steadfast interior thermal conditions orchestrating minimal exposure to direct sunshine and release of heat at night. The second prototype uses a small, clustered design which limits the exposure of the envelope but has functional internal layouts to fit into limited urban plots.

As shown in Fig. 8, the principles of geometry-led houses can be translated into easily constructible housing typologies that can be adapted to the practice of

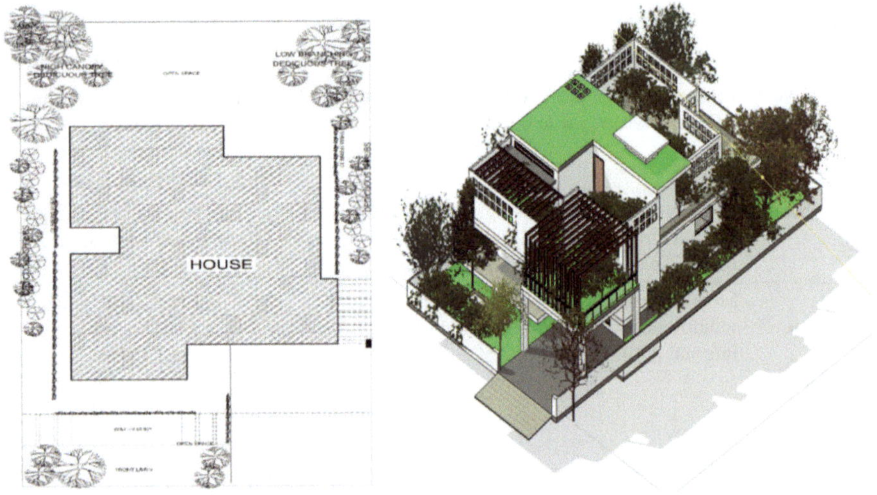

Fig. 8 Optimised prototype design. *Source* Authors

regular plot sizes and development. These prototypes illustrate that major initiatives in reducing cooling dependency can be made by spatial configuration only, which confirms the practicability of implementing the resilience-oriented geometry in typical residential design and building procedures.

7 Implications for Climate-Resilient Urban Design

Although geometry-based approaches to buildings contribute greatly to the thermal performance of buildings and the minimisation of cooling loads, their resilience potential can only be fully unlocked when those principles are integrated into the increased scope of urban form and regulation. The morphology of neighbourhoods, the direction of streets and the distribution of open spaces all play a role in affecting ventilation networks, shading each other and storing heat cumulatively between residential districts. Even well-planned individual dwellings, when poorly coordinated with urban form adjacent to them, will be compromised by surrounding urban form that inhibits airflow and traps long-wave radiation. This can be done through neighbourhood designs in hot semi-arid cities like Lahore in which the density is limited but still allows ventilation access and shaded walk paths, which help reduce ambient temperatures and enhance the comfort of outdoor environments. Minimised roof exposure, spaced out courtyards, and height to width ratios in favour of airflow all moderate the microclimates and contribute to the efficacy of building-level passive strategies. Such spatial synergies prove that geometry plays the role of passive urban infrastructure, which helps to reduce heat without even using more energy. Governance wise, there are more frameworks that govern development control within the

Table 5 Guidelines for various stakeholders

Stakeholder	Priority Actions	Resilience Outcome
Architects	Orientation control, courtyard integration, façade shading	Lower cooling loads, thermal comfort
Planners	Form-based zoning, open-space ratios	Urban heat mitigation
Policymakers	Building performance standards, incentives	Scalable climate adaptation
Communities	Shading, vegetation, retrofits	Household heat protection

existing systems, which tend to focus on plot efficiency, larger area that can be built, at the expense of climatic responsiveness. A redesign of building by-laws to include form-based control including orientation-sensitive layout, minimum open-space requirements and height to width ratios would enable climate adaptation aims to be realised by the normal execution of planning approvals. City-scale impact is only possible with such regulatory alignment, especially in situations which are rapidly urbanising and in which cumulative housing decisions determine long-term thermal vulnerability. The geometry-led climate adaptation needs to be effectively implemented through the coordinated involvement of stakeholders of housing production and regulation. The architects affect early-stage spatial decisions, planners affect morphology of neighbourhoods by use of zoning and layout control, policymakers are setting performance criteria and incentives, researchers are offering evidence that is locally based, and communities are assisting in providing the maintenance of common microclimatic resources such as green spaces and shaded streets. These roles can be aligned to mainstream form-based strategies into housing delivery instead of limiting it to experimental or high-end developments.

Practical measures in all three areas, i.e., professional, institutional, and community, are outlined in Table 5, which adds to the fact that climate-resilient housing must be based on integrated governance and not on technical micro-solutions. Geometry-led strategies can be deployed at scale by reconciling design practice and regulatory regimes, as well as by enlisting the involvement of the public to achieve the goals of thermal safety, lowering peak electricity demand, and enhancing adaptive capacity to frequent heat extremes by other means.

8 Conclusions

This chapter demonstrates that geometry-led residential design is a critical strategy for strengthening thermal and energy resilience in hot semi-arid urban contexts such as Lahore. The findings confirm that early-stage spatial decisions including building orientation, compact massing, courtyard integration, and façade shading significantly influence cooling demand and indoor thermal stability. Architectural geometry is therefore shown to be a key determinant of a building's vulnerability to heat stress, rather than merely a formal or stylistic choice. By emphasising form-based interventions over technology-dependent solutions, the study positions geometric design as a low-energy and economically accessible mechanism for climate adaptation. These strategies operate continuously, reduce reliance on mechanical cooling, and enhance a building's capacity to maintain thermal comfort during extreme heat events and periods of energy stress, thereby mitigating overheating risks and associated health impacts. When applied at the scale of residential developments, geometry-led strategies also contribute to reducing neighbourhood-level heat accumulation and peak electricity demand, reinforcing urban energy resilience. The translation of simulation-based findings into design principles and planning implications demonstrates that climate-resilient geometry can be systematically embedded within architectural practice and regulatory frameworks. This scalability makes geometry-led housing particularly relevant for rapidly urbanising cities in the Global South, where reducing cooling dependency and enhancing heat resilience are increasingly urgent under future climate conditions.

References

1. Oke TR (1982) The energetic basis of the urban heat island. Q J R Meteorol Soc 108(455):1–24
2. Santamouris M (2015) Regulating the damaged thermostat of the cities status, impacts and mitigation challenges. Energy Build 91:43–56
3. Givoni B (1998) Climate considerations in building and urban design. Van Nostrand Reinhold, New York
4. Ratti C, Baker N, Steemers K (2005) Energy consumption and urban texture. Energy Build 37(7):762–776. https://doi.org/10.1016/j.enbuild.2004.10.010
5. Sharifi A, Yamagata Y (2016) Urban resilience assessment: multiple dimensions and challenges. Sustain Cities Soc 26:1–17. https://doi.org/10.1016/j.scs.2016.05.009
6. IPCC (2021) Sixth assessment report: Working Group I The physical science basis. Intergovernmental Panel on Climate Change
7. Aamer F (2024, October 31) Building sustainable futures: advancing climate resilience in South Asia. Asia Society Policy Institute in collaboration with World Bank South Asia Region
8. UN-Habitat (2020) World cities report: the value of sustainable urbanization. United Nations Human Settlements Programme
9. Vardoulakis S, Dear K, Hajat S, Heaviside C, Eggen B, McMichael AJ (2015) Comparative assessment of the effects of climate change on heat- and cold-related mortality in urban areas. Environ Health 14(Supp 1):S3. https://doi.org/10.1186/1476-069X-14-S1-S3
10. Pakistan Meteorological Department (2020) Annual climate data for Lahore. Government of Pakistan
11. Emmanuel R, Johansson E (2006) Influence of urban morphology and sea breeze on hot humid microclimate: the case of Colombo, Sri Lanka. Clim Res 30(3):189–200
12. Ali-Toudert F, Mayer H (2006) Numerical study on the effects of aspect ratio and orientation of urban street canyons on outdoor thermal comfort in hot and dry climate. Build Environ 41(2):94–108. https://doi.org/10.1016/j.buildenv.2005.01.013
13. Kahoorzadeh A, Shahwarzi S, Farjami E, Osivand S (2014) Investigation of passive solar strategies in residential buildings. Int J Environ Sci Dev 5(2):132–136. https://doi.org/10.7763/IJESD.2014.V5.464

14. Olgyay V (1963) Design with climate: Bioclimatic approach to architectural regionalism. Princeton University Press
15. Attia S, Hensen J, Beltrán L, De Herde A (2013) Selection criteria for building performance simulation tools: contrasting architects' and engineers' needs. J Build Perform Simul 6(3):155–169. https://doi.org/10.1080/19401493.2012.693059

Aqua Vascular Tower: A Biomimetic Approach to Climate-Resilient Vertical Architecture in Dense Urban Settings

Sadia-E-Jannat, Md Imamul Alam Ishat, and Nadia Monzur

1 Introduction

Urbanization in the twenty first century is moving faster than ever. Cities are expanding upwards, but this also makes them more vulnerable to climate risks. Densely populated megacities like Singapore face serious issues, including limited land, high energy use, higher urban heat, and increased risk of extreme weather. In this situation, architecture needs to change. It must shift from being static structures to dynamic systems that blend environmental, social, and functional intelligence.

This chapter presents the Aqua Vascular Tower, a conceptual high-rise design that rethinks the skyscraper as a living, circulatory ecosystem. Inspired by the human cardiovascular system, the tower is designed as a layered structure that responds to climate stress, manages water and air flows, and improves occupant well-being. Located in Singapore, a city-state known for its density, innovation, and commitment to green urban spaces, the Aqua Vascular Tower represents an architectural model that imitates biological systems in both shape and function.

The chapter aims to connect biomimetic design theory with real-world architectural applications for climate-resilient high-rises. It describes the process of creating and simulating the tower's unique façade system, climate strategies, and social integration methods. Using parametric modeling tools like Rhino and Grasshopper, along with Autodesk Flow Design for environmental simulation, the design examines how changing façade shapes and adding hydroponic gardens can improve both sustainability and spatial quality.

Sadia-E-Jannat (✉) · N. Monzur
Department of Architecture, Rajshahi University of Engineering & Technology, Rajshahi, Bangladesh
e-mail: sadiae.jannat29@gmail.com

M. I. A. Ishat
Department of Mechanical Engineering, Chittagong University of Engineering & Technology, Chittagong, Bangladesh

A. Cheshmehzangi et al. (eds.), *Design Innovation for Climate-Resilient Cities and Built Environments*, Urban Sustainability, https://doi.org/10.1007/978-981-92-2297-1_5

In the end, the Aqua Vascular Tower envisions the future of vertical urban living as more than just a place to reside. It presents a regenerative system that gathers rainwater, cools the air, absorbs carbon, and supports vertical biodiversity. This chapter argues that design driven by biomimicry is crucial for rethinking architecture's role in creating resilient cities.

2 Contextual Background: Singapore as a Living Lab

Singapore, despite its small size, is one of the most urbanized countries globally, with a population density of over 8000 people per square kilometer [1]. The city-state is recognized for its careful city planning, environmental management, and global leadership in vertical greening and sustainable infrastructure. However, it still faces significant environmental challenges, such as limited land, rising temperatures, high energy use, and more intense rainfall due to climate change. This situation makes Singapore a unique location for testing innovative urban architecture.

In terms of climate, Singapore has a tropical rainforest climate (Köppen classification: Af). It has high humidity, ranging from 75 to 95%, consistent sunlight, and yearly rainfall averaging 2400 mm. These conditions increase the need for combined design strategies, especially for water management, heat reduction, and optimizing natural light. While traditional high-rise buildings often overlook these factors for standardization and to maximize surface area, new vertical design models should work like ecological systems that are responsive, flexible, and efficient [2].

Urban projects like Marina One, Oasia Hotel Downtown, and CapitaSpring demonstrate the benefits of vertical gardens, sky terraces, and integrated landscapes [3, 4]. These features help lower temperatures, improve air quality, and increase urban biodiversity. However, most of these projects still use standard façades with added greenery. The Aqua Vascular Tower goes further by integrating environmental performance into the shape and function of the façade, using biomimicry as the main design principle [2].

The Singaporean government supports sustainability and smart city initiatives, encouraging such innovation. With substantial public and private investments in green building certifications like the BCA Green Mark and urban biodiversity efforts such as the City in Nature vision, the city promotes architectural experimentation that addresses global urban challenges locally. Additionally, the city's frequent rain and vertical development provide a great chance to explore passive rainwater harvesting systems in high-rise façades.

By placing the Aqua Vascular Tower in Singapore, the design takes advantage of the city's climate, density, and supportive policies to showcase the practicality of biomimetic vertical architecture. This context not only clarifies the tower's environmental and design rationale but also connects the project to Singapore's broader goals for urban resilience.

3 Biomimicry in Architecture: Theoretical Foundations

Biomimicry is the practice of using nature's systems, processes, and forms as inspiration for design. This performance-oriented approach has been increasingly explored in biomimetic architectural research and environmentally responsive building envelopes [5, 6]. It provides a strong way to rethink how buildings can better respond to environmental challenges. Unlike traditional architecture that forces solutions onto nature, biomimetic design aims to imitate nature's efficiency, adaptability, and resilience. In a world that is rapidly urbanizing and facing growing climate issues, this approach not only supports sustainability but also opens up new possibilities for architectural design.

The roots of biomimicry reach across different fields. They range from Leonardo da Vinci's studies of bird flight to modern uses in product design, robotics, and architecture. Janine Benyus, in her important book Biomimicry: Innovation Inspired by Nature [7], established biomimicry as a method for sustainable design, stressing that "life creates conditions conducive to life." In architecture, this means creating buildings that function like living organisms, systems that adapt, self-regulate, and interact with their surroundings.

Examples of biomimicry in architecture include the Eastgate Centre in Harare, Zimbabwe, designed by Mick Pearce. It takes inspiration from termite mounds to create a passive cooling system. Another example is the Eden Project in the UK, where the geodesic domes mimic soap bubbles to optimize surface area and structural efficiency. These cases show how strategies inspired by nature can lead to advancements in thermal comfort, structural efficiency, and resource use.

The Aqua Vascular Tower continues this tradition by using the human cardiovascular system as its guiding concept. This system is one of the most complex and efficient in nature. It transports oxygen, nutrients, and waste through a detailed network of veins, arteries, and capillaries. In living organisms, the cardiovascular system reacts to environmental changes, adjusting blood flow to manage temperature, oxygen levels, and hydration. This ability makes it a valuable model for designing building systems that control water, airflow, and thermal conditions in a tall structure.

In the tower's design, this biological idea comes to life through a façade system that acts like vascular tissue. It collects and distributes rainwater, encourages vertical airflow, and promotes passive cooling. The hydroponic gardens in the façade use capillary action to move water upward, nourishing plants without needing mechanical pumps. The building's shape, created with parametric tools, reflects the branching patterns of blood vessels, ensuring that environmental resources are effectively distributed across its height.

By integrating performance into the building's form, the Aqua Vascular Tower advances the concept of architecture as metabolism. Here, structural, climatic, and ecological systems are deeply connected instead of being added later as an afterthought. This nature-inspired approach not only improves environmental performance but also reimagines the skyscraper as a living organism within the urban landscape.

By connecting architectural thinking with biological principles, biomimicry provides a foundation for innovation that unites analytical reasoning with design creativity. The Aqua Vascular Tower applies this approach within a dense urban context, demonstrating how bio-inspired strategies can generate high-rise forms capable of adapting to environmental and climatic demands.

Key Takeaways

- Biomimicry offers a design framework that prioritizes adaptability, efficiency, and resilience by learning from natural systems.
- In architecture, biomimetic principles enable buildings to behave more like living organisms, self-regulating and environmentally responsive.
- Precedents such as termite-mound-inspired ventilation and biologically optimized structures validate biomimicry's performance potential.
- The human cardiovascular system provides a powerful analogy for managing water, airflow, and thermal regulation in tall buildings.
- The Aqua Vascular Tower integrates form, structure, and environmental performance into a unified metabolic system rather than applied technologies.

4 Design Methodology and Concept Development

The Aqua Vascular Tower was developed through an iterative process that combined biomimetic inspiration, computational modeling, and environmental simulation. Unlike conventional workflows in which form typically precedes performance, this project adopted a performance-first methodology. The design began with climate data and biological analogies, ensuring that form, function, and environmental responsiveness were integrated from the outset. Each stage of the process, from massing to façade articulation, was grounded in the cardiovascular metaphor, interpreted through parametric tools and climatic logic.

4.1 Conceptual Genesis: From Vessel to Vertical System

The design process began with an analysis of the hierarchical branching patterns of human arteries and veins, which informed the concept of a vertically circulatory façade. Analogous to the cardiovascular system, where arteries transport oxygenated blood from the heart to all parts of the body, the tower's façade was envisioned as a network that collects and redistributes rainwater, airflow, and daylight across its vertical surface. This vascular logic directly shaped the building's form, resulting in a branching vertical core with secondary "veins" articulated across the envelope (Fig. 1).

Importantly, the biological metaphor extended beyond aesthetics. It influenced spatial organization, facade geometry, and internal circulation strategies. Rather than

Fig. 1 Idea generation. *Source* Diagram illustrated by authors

being conceived as a monolithic volume, the tower functions as a multi-layered system in which each architectural component, structural ribs, hydroponic gardens, water channels, and shading devices, operates as part of an integrated organism.

Key Takeaways

- The human cardiovascular system provides a functional model for vertical resource distribution in high-rise architecture.
- Vascular logic influences not only façade aesthetics but also spatial hierarchy, circulation, and environmental systems.
- The tower is conceived as a **multi-layered organism**, where structural, ecological, and climatic elements operate interdependently.
- Biomimicry is applied as a **systemic framework**, not a symbolic or decorative reference.

4.2 Environmental Simulation and Aerodynamic Optimization

Following the exploration inspired by the cardiovascular system, the tower's shape was refined through aerodynamic studies tailored to Singapore's climate. The initial designs started with a rectangular or square base. They were then rotated by 45

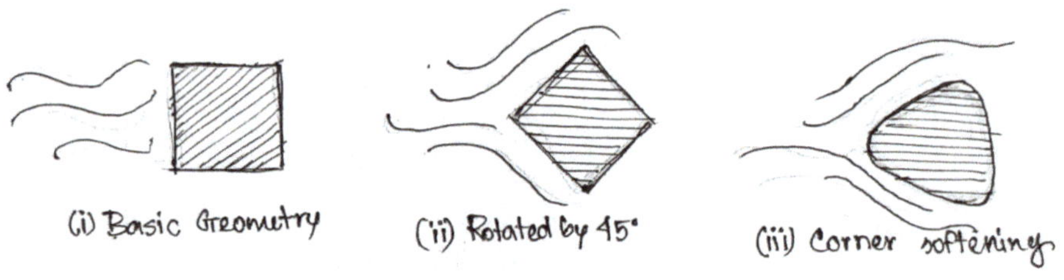

Fig. 2 Building orientation. *Source* Diagram illustrated by authors

degrees to face the prevailing wind direction. This rotation helped reduce direct wind pressure and improve natural ventilation (Fig. 2).

Additional changes, such as rounding the corners and creating tapered profiles, were made to lessen vortex shedding and distribute wind loads more evenly along the vertical surface (Fig. 3).

This sequence shows how the design of a high-rise tower changes through three geometric transformations, each aimed at improving aerodynamic performance. The first form is a simple vertical extrusion with a triangular shape, reflecting a basic and stiff structure. In the second stage, the tower narrows as it rises to reduce wind pressure and drag forces. The final stage adds a helical twist, resulting in a lively and smooth shape that helps disrupt wind flow patterns. This twist can reduce vortex shedding and enhance overall stability. Besides its aerodynamic benefits, the twisted shape also adds to the tower's visual appeal, creating a sense of movement and elegance while responding effectively to environmental forces.

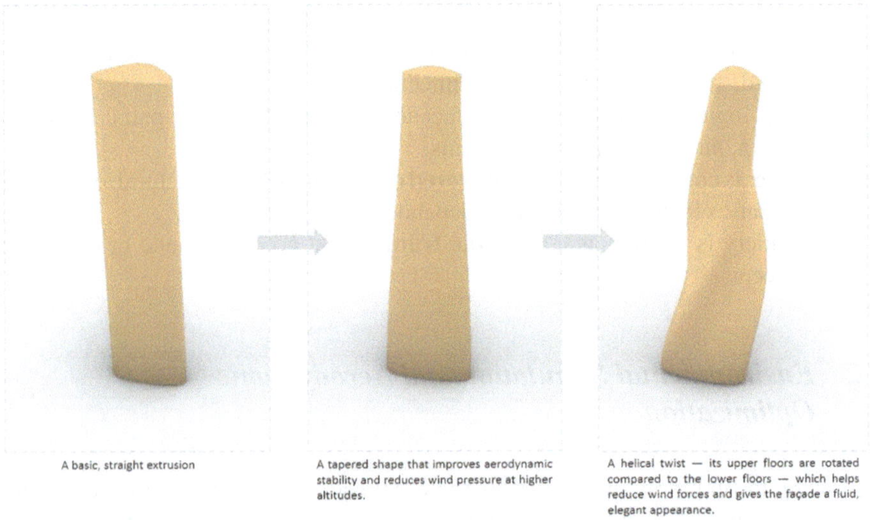

Fig. 3 Conceptual evolution of tower form for aerodynamic enhancement. *Source* Diagram illustrated by authors

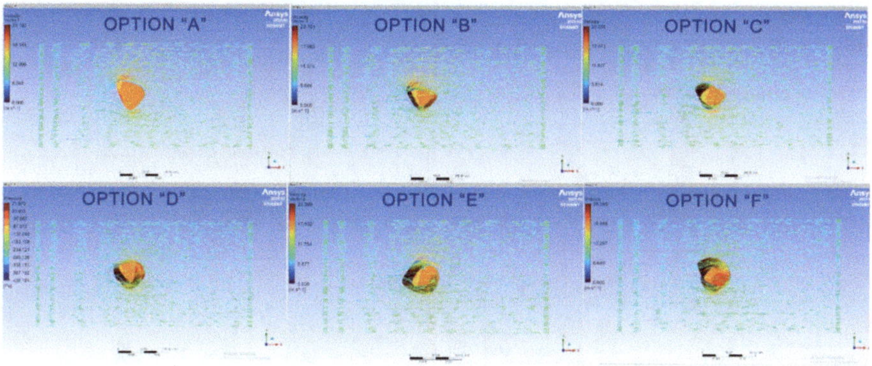

Fig. 4 ANSYS simulation results comparing six tower forms (**A–F**) at 10 m/s wind velocity: (a) velocity vectors. *Source* Simulated by authors in ANSYS Student 2025R2

To assess aerodynamic performance, six different tower shapes (A–F) were tested using ANSYS Student Software with a steady inlet speed of 10 m/s. This speed reflects the highest average wind speed recorded during the Northeast Monsoon season (January–February). Each shape was evaluated based on pressure contours, velocity vectors, and surface flow distribution to analyze drag characteristics, turbulence intensity, and wake formation.

Velocity vector plots (Fig. 4) illustrating airflow behavior around six tower forms (A–F) at an inlet velocity of 10 m/s. The visualization reveals how each geometry influences wind deflection, turbulence intensity, and wake formation. Streamline continuity and vortex size indicate the relative aerodynamic efficiency and stability of the tested configurations.

Pressure contour maps (Fig. 5) showing wind pressure distribution across the façades of six tower forms. The variation between high-pressure (red) and low-pressure (blue) zones demonstrates how geometry affects drag forces and suction effects on the leeward side. This comparison helps evaluate wind load distribution and potential aerodynamic performance.

Surface pressure distribution (Fig. 6) illustrating airflow interaction along tower façades. The gradient of velocity magnitudes highlights regions of flow attachment, separation, and recirculation. By comparing these patterns among forms A–F, aerodynamic characteristics such as drag reduction, turbulence minimization, and overall wind comfort potential are clearly visualized.

Table 1 summarizes the comparative findings of the six tested configurations.

The simulation results (Figs. 4–6) showed that Forms E and F had the best aerodynamic traits, with smaller wake areas and lower pressure differences. However, Form E was chosen for further development because it combined aerodynamic efficiency with architectural functionality better than the others. It allowed for smooth airflow and reduced turbulence while keeping a more efficient floor plate ratio for mixed-use planning. The aerodynamic design leads to lower lateral wind loads, better

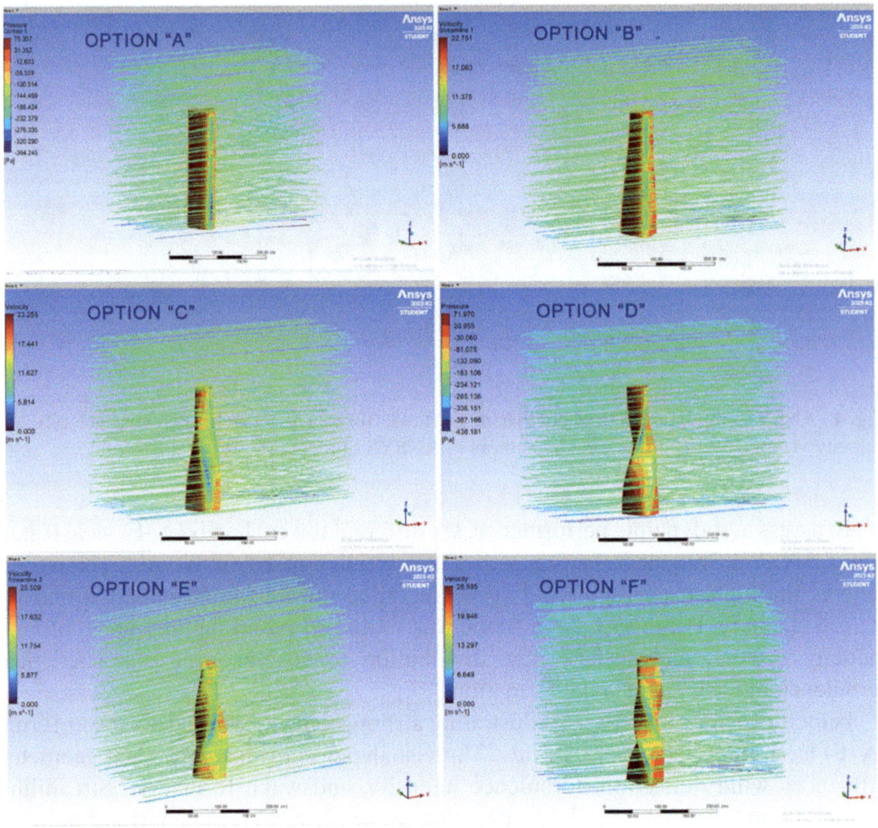

Fig. 5 ANSYS simulation results comparing six tower forms (**A–F**) at 10 m/s wind velocity: (b) pressure contours. *Source* Simulated by authors in ANSYS Student 2025R2

wind comfort, and fits well with sustainable vertical urban development strategies in crowded areas.

Key Takeaways

- Tower orientation and geometric transformations are optimized based on Singapore's prevailing wind conditions.
- Aerodynamic simulations using ANSYS evaluated six massing variants under a 10 m/s inlet wind velocity.
- Streamlined and tapered geometries significantly reduce turbulence, wake formation, and pressure differentials.
- Form E was selected as the optimal solution, balancing aerodynamic performance with usable floor area.
- Simulation-driven massing contributes to reduced wind loads, improved pedestrian comfort, and structural efficiency.

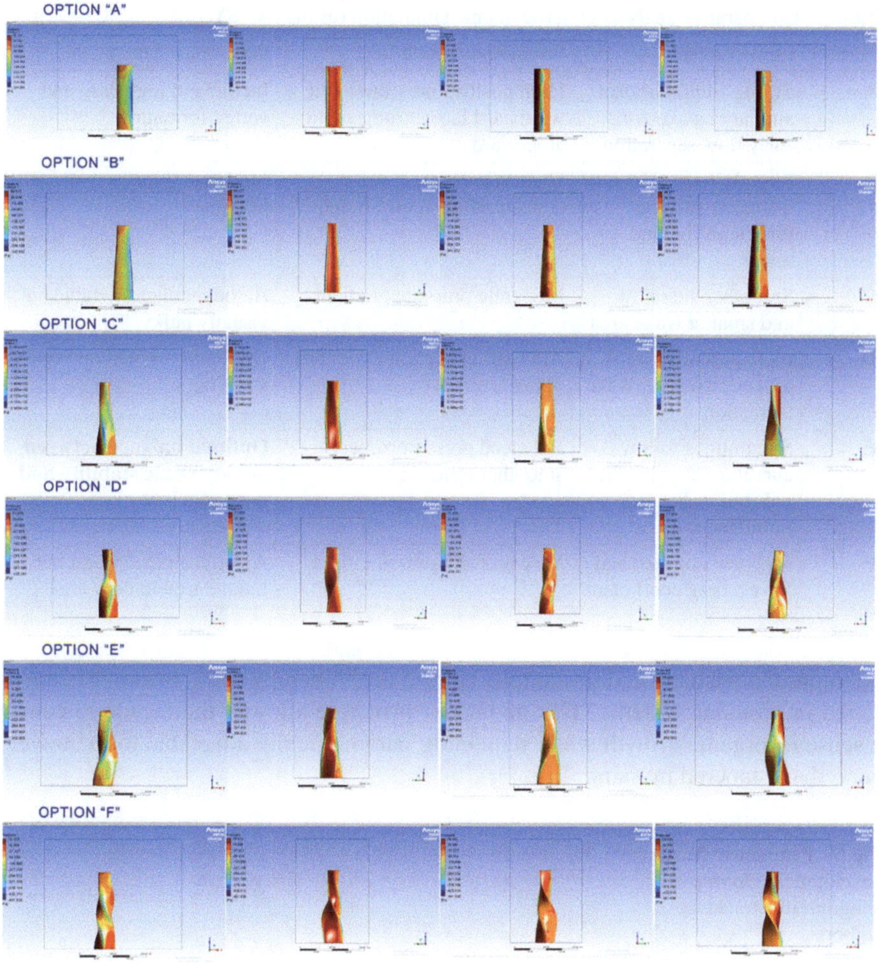

Fig. 6 ANSYS simulation results comparing six tower forms (**A**–**F**) at 10 m/s wind velocity: (c) surface pressure distribution. *Source* Simulated by authors in ANSYS Student 2025R2

4.3 Parametric Facade Development and Environmental Integration

The facade concept of the Aqua Vascular Tower was generated through a parametric design workflow using Rhinoceros 3D and Grasshopper. At this conceptual stage, the parametric system functioned primarily as a form-finding and organizational tool, translating the biomimetic inspiration of the human cardiovascular system into a buildable architectural geometry.

A Voronoi algorithm was applied within Grasshopper to develop the façade's patterning logic. The Voronoi tessellation produced a cellular network of irregular

Table 1 Comparative analysis of aerodynamic simulation results (ANSYS)

Form	Flow pattern	Pressure distribution	Performance comment
A	Strong frontal impact and large wake zone; early flow separation	High positive pressure at the windward face; strong suction at leeward	Inefficient; high drag and vortex formation
B	Slightly improved side diffusion but large turbulence behind tower	Uneven pressure concentration at mid-height	Moderate improvement; still unstable
C	Smoother side flow and smaller wake area	Moderate pressure gradient	Aerodynamically stable but visually bulky
D	Flow partially wraps around; irregular wake formation	Alternating suction zones near upper levels	Oscillatory; potential vortex shedding
E	Streamlines follow curvature effectively; reduced wake and turbulence	Balanced pressure zones; smoother transition	**Optimal balance between aerodynamic stability and floor efficiency**
F	Most streamlined and lowest drag coefficient	Very uniform pressure field	Aerodynamically best but limited usable floor area

polygons that visually and structurally echoed the branching and connective behavior found in vascular systems. This geometric principle allowed the façade to convey a sense of organic growth while remaining parametrically adaptable to the tower's curved and tapered massing (Fig. 7).

Fig. 7 Parametric facade development. *Source* Diagram illustrated by authors

The primary goal of this stage was not to achieve exact environmental optimization but to explore a performative architectural language capable of integrating multiple potential functions within a single system. The resulting façade demonstrates how a biomimetic pattern can conceptually support environmental responsiveness through its depth, porosity, and hierarchical structure. Although detailed simulations were not yet conducted, the Voronoi framework suggests several latent performance potentials that can be further developed in subsequent research phases:

- Reduction of lateral wind pressure through partial permeability and distributed surface geometry;
- Opportunities for shading and daylight modulation, derived from the pattern's varying cell densities;
- Integration of rainwater collection paths along the cellular joints, allowing gravitational flow toward lower collection channels;
- Potential structural efficiency, if the pattern were rationalized into a triangulated or lattice-support system.

Thus, the facade operates as a conceptual climatic membrane, merging the poetic logic of natural systems with the algorithmic precision of digital tools. It illustrates how parametric methods can be used not merely to generate form but to speculate on environmental performance, even in the absence of complete analytical validation.

The Voronoi-based facade remains a prototype stage in the design process, symbolic of the tower's vascular metaphor and ready to evolve through future integration with structural, environmental, and hydraulic simulations.

Key Takeaways

- Parametric modeling enables iterative testing of façade porosity, circulation density, and environmental performance.
- CFD and environmental simulations provide indicative insights into airflow, cooling potential, and system interaction.
- All simulations are based on realistic climatic data and established assumptions suitable for early-stage design research.
- The methodology emphasizes **relative performance trends**, ensuring transparency and reproducibility without overclaiming precision.

4.4 *Integrated Hydroponic Strategy*

A central sustainable feature of the design is its hydroponic vertical garden system, integrated into the facade. Unlike soil-based systems, hydroponics reduces weight, minimizes maintenance, and optimizes water use, particularly suited to Singapore's humid tropical climate. Key features include:

- Modular hydroponic pods anchored to vascular tubes, mimicking capillary action
- Gravity-fed drip systems distributing rainwater collected by the facade

- Plant selection focused on local species with high evapotranspiration potential (e.g., ferns, philodendrons, pothos)

This system not only supports passive cooling and biodiversity but also aligns with Singapore's City in Nature vision, promoting ecological integration in dense vertical contexts.

Key Takeaways

- Hydroponic systems offer a lightweight and water-efficient alternative to soil-based vertical greenery. Similar findings have been reported in studies on vertical farming and hydroponic cultivation systems [1, 8].
- Gravity-fed and modular hydroponic components reduce mechanical dependence and maintenance complexity.
- Plant selection emphasizes local species that enhance evapotranspiration, cooling, and biodiversity.
- The system supports passive microclimate regulation while aligning with Singapore's City in Nature agenda.
- Integrated vertical greenery strengthens the tower's role as a climate-responsive and ecologically active high-rise.

4.5 Methodological Assumptions and Simulation Parameters

The environmental performance evaluation of the Aqua Vascular Tower was conducted using a **conceptual yet replicable simulation framework**. Aerodynamic behavior was assessed using ANSYS Student software under steady-state conditions, with an inlet wind velocity of **10 m/s**, corresponding to peak seasonal average wind speeds in Singapore. Comparative analyses were performed across six tower geometries to evaluate pressure distribution, wake formation, and airflow continuity, assuming neutral atmospheric stability and smooth surface conditions to isolate geometric effects.

Rainwater harvesting potential was estimated using standard hydrological equations based on **annual rainfall data (2400 mm)**, an approximated façade collection area derived from massing geometry, and a **runoff coefficient of 0.7**, reflecting smooth, channelized vertical surfaces. These assumptions follow established methods used in early-stage building-scale water system assessments [8]. While the simulations do not represent full operational performance, they provide a transparent and reproducible baseline for comparing form-driven environmental behavior and informing future high-resolution analyses.

5 Facade System and Environmental Strategies

The Aqua Vascular Tower's facade is the main architectural feature where bio-inspired design meets environmental performance. It draws inspiration from the human vascular system. The facade does more than just protect the building; it creates a circulatory link between the structure and its surroundings. This system collects and redistributes rainwater, supports passive cooling, improves natural ventilation, maximizes daylight, and allows for hydroponic planting, all while keeping a unique architectural style.

5.1 Vascular Facade Logic: Circulation of Water, Air, and Light

The tower's external skin is composed of a series of vertically oriented vascular tubes, branching out from a central core and wrapping the façade in dynamic loops. These tubes serve multiple functions:

- Rainwater channels: Direct rainfall downward into a gravity-fed collection system
- Airflow conduits: Promote cross-ventilation through low-pressure zones and porous openings
- Daylight filters: Use operable shading membranes and perforations to modulate solar gain
- Structural ribs: Provide lateral bracing while housing hydroponic modules

Each vascular path is parametrically adjusted based on its height, orientation, and exposure, forming a responsive skin that changes across the tower's height, much like how arteries vary in diameter and elasticity throughout the human body.

5.2 Rainwater Harvesting System: Estimation and Performance

Singapore receives an average of 2400 mm of rainfall annually, making it one of the world's wettest urban environments. The Aqua Vascular Tower takes advantage of this climatic condition through a vertically integrated rainwater harvesting façade. The conceptual estimation of harvesting potential follows the basic hydrological formula:

$$\text{Annual Collection (liters)} = \text{Rainfall (mm)} \times \text{Collection Area (m}^2\text{)} \times \text{Runoff Coefficient}$$

For this stage of the research, the façade surface area has been approximated geometrically based on the conceptual tower massing. The tower's average height and perimeter yield an estimated vertical collection area of approximately 12,000 m^2 (based on an average of 3000 m^2 per façade). A runoff coefficient of 0.7 is used to account for smooth, channelized vertical surfaces that facilitate gravity-fed flow.

Calculation:

$$2400\ \text{mm} = 2.4\ \text{m}$$

$$\text{Total Area} = 3000\ \text{m}^2 \times 4 = 12{,}000\ \text{m}^2$$

$$\text{Annual Collection} = 2.4 \times 12{,}000 \times 0.7 = 20{,}160{,}000\ \text{L/year} \approx 55{,}223\ \text{L/day}$$

This harvested water can be stored in integrated cisterns at intermediate levels and used for:

- Irrigating hydroponic gardens
- Cooling mist systems in terraces
- Non-potable graywater systems

Rainwater harvesting remains one of the most effective passive strategies for reducing potable water demand in buildings and enhancing urban water resilience [9]. The system minimizes the need for mechanical pumps, relying instead on gravity-fed flow through the tower's branching vascular network, where collected water descends from upper modules to lower reservoirs.

It should be noted that, at this conceptual stage, the façade area and runoff estimations are approximate. Detailed parametric analysis using Rhinoceros and Grasshopper will be conducted in future research to refine facade area measurements and validate hydrological performance. Nonetheless, the current estimation provides a credible baseline to illustrate the tower's potential as a self-sustaining vertical water system in a tropical urban environment. Comparable facade-integrated environmental systems have also been investigated in hydroponic and climate-responsive building envelopes [10].

5.3 Passive Cooling and Thermal Comfort

With Singapore's high temperatures (27–32 °C) and humidity (75–95%), we focused on passive cooling methods:

- Vertical gardens provide evaporative cooling, lowering the façade's temperature by 2–3 °C.
- Shading elements in the vascular ribs block direct sunlight while allowing diffuse light in.
- Cross-ventilation zones between living units and green buffers act as thermal buffers.

- Wind tunnels carved into mid-tower voids increase airflow, creating pressure differences that promote natural ventilation.

Flow simulations showed that the tower's curved shape enhances stack effect ventilation, improving air circulation without mechanical support.

5.4 Natural Daylighting and Solar Control

The building envelope was designed to balance solar access and shading across varying heights and orientations:

- South-facing façades use deeper vascular channels as self-shading overhangs.
- East and west façades include adaptive louvers inspired by venous valves, responsive to solar angles.
- North-facing modules prioritize daylight penetration with higher window-to-wall ratios.

The vascular system's organic curvature diffuses light, reducing glare while maintaining brightness. Light shelves embedded into horizontal tubes reflect light deeper into interior spaces, reducing dependence on artificial lighting.

Key Takeaways

- The façade functions as an **active environmental system**, integrating water, air, light, and vegetation through a biomimetic vascular logic.
- Rainwater harvesting and gravity-fed circulation demonstrate the potential for **low-energy water reuse** in tropical high-rise buildings.
- Passive cooling strategies, vertical greenery, aerodynamic form, and natural ventilation, enhance thermal comfort while reducing mechanical dependence.
- Parametric design enables façade adaptability to orientation and climate, balancing **solar control and daylight access**.
- The chapter illustrates how **biomimicry can translate environmental performance into architectural form**, supporting climate-resilient vertical urbanism.

6 Social and Urban Integration

While the Aqua Vascular Tower focuses on biomimetic and environmental strategies, it also emphasizes social sustainability. In crowded cities like Singapore, vertical buildings should do more than respond to environmental factors. They need to revive community life, promote well-being, and include public spaces in their design. This section looks at how the tower design addresses the social aspects of high-rise living through accessible green features, mental health strategies, and inclusive spaces.

6.1 *Vertical Public Realms and Communal Terraces*

In most traditional towers, social interactions happen only at ground level. The Aqua Vascular Tower changes this by bringing public spaces upwards, incorporating multi-level terraces, sky gardens, and shared platforms distributed throughout the structure. These platforms serve as:

- Community gardens for residents and schools.
- Quiet retreats for relaxation, meditation, and solitude.
- Fitness and therapy spaces meant for movement, rehabilitation, and wellness.
- Public viewing areas that provide panoramic views of the city.

The vascular façade system surrounds these terraces, making them feel both enclosed and open. This design keeps a connection to nature and climate while ensuring comfort.

By stacking semi-public and communal functions vertically, the tower restores the social aspects often missing in high-rise living. This approach encourages casual interactions, urban farming, and connections across generations in a space-efficient way.

6.2 *Water Cycle and Adaptive Hydrology*

The Aqua Vascular Tower acts as a combined water system, creating a closed-loop water cycle that mimics the adaptability of natural ecosystems. In dense tropical areas like Singapore, which receives over 2400 mm of rain each year, managing water is both necessary and a chance to build climate resilience. The tower's design features a multi-layered water system that collects, filters, stores, and redistributes rainwater throughout its height, allowing for adaptable hydrology in a tall structure.

Rainwater is captured through the facade's channels (Fig. 8) and guided by gravity into filtration chambers. These chambers use natural filtering materials and light gravel to remove impurities before directing the water into storage tanks located within mechanical floors. From there, the recycled water is distributed for three main purposes:

- Hydroponic irrigation for the vertical gardens.
- Cooling systems through misting and evaporative panels found on terraces.
- Non-drinking uses like toilet flushing and maintenance.

This closed-loop system works with minimal mechanical energy, depending mainly on gravity and capillary action. It shows the idea of adaptive circulation, where water continuously flows through the system based on environmental factors such as rainfall, humidity, and temperature.

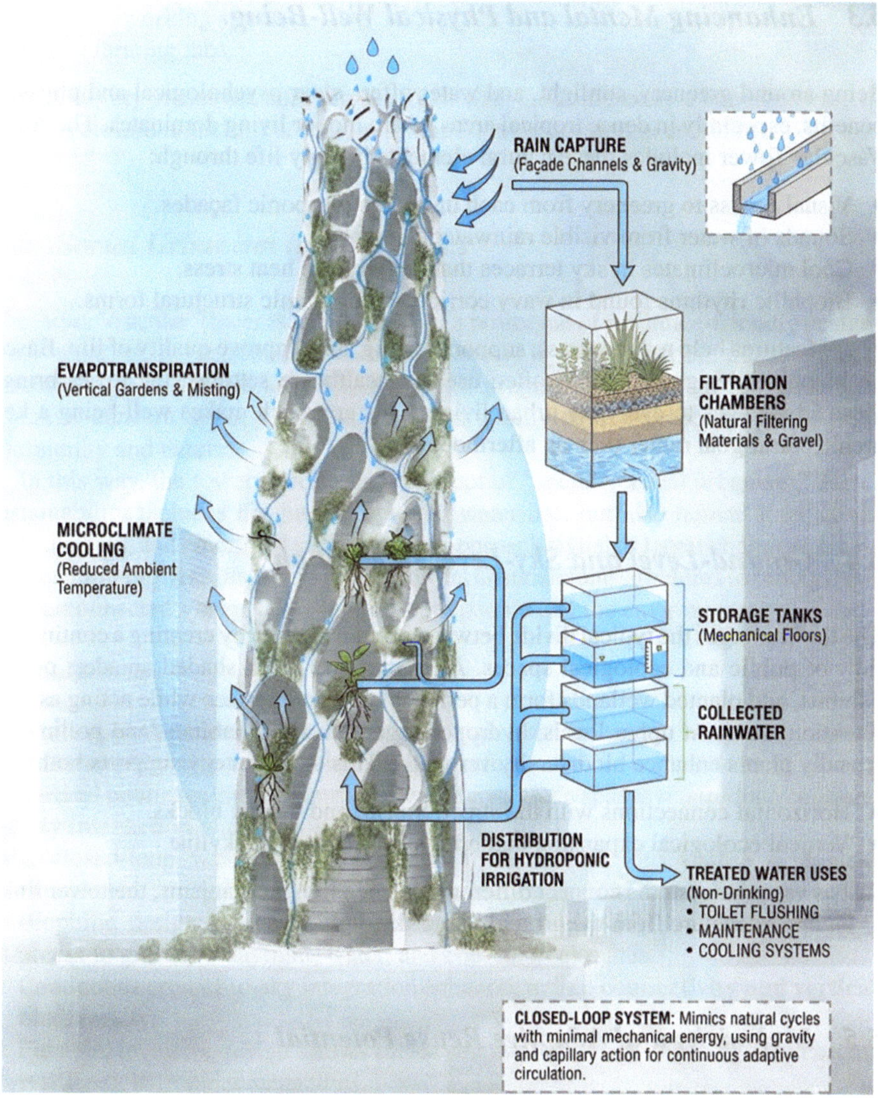

Fig. 8 Rainwater harvesting and circulation system of the aqua vascular tower. *Source* Diagram illustrated by authors

By turning a facade into a responsive water surface, the Aqua Vascular Tower changes traditional water management from a background service into visible ecological infrastructure. The ongoing movement of water through the tower not only provides environmental benefits but also creates a sensory and aesthetic link to nature. This connection raises occupants' awareness of resource cycles and strengthens the tower's identity as a living, climate-adaptive building.

7.1 From Speculation to Realization: Technological Gaps

Although the project is feasible and possible with current technology, several technical and material limitations are holding back its full implementation:

- Creating vascular façades with complex fluid dynamics and integrated hydroponic systems needs multi-material 3D printing or modular construction, which are not widely used on a large scale.
- Monitoring and controlling fluid flow for rainwater and airflow on a dynamic vertical surface requires smart sensors and automated valves. These technologies are still developing in the building industry.
- Managing high-rise hydroponic gardens on curved surfaces poses logistical challenges for accessibility, irrigation, and plant replacement.

These issues highlight the gap between ideas inspired by nature and what can actually be built. They point out the need for collaboration among architects, engineers, biologists, and computational designers.

7.2 Feasibility in Tropical Megacities

Singapore has strong governance, clear sustainability goals, and a solid technological infrastructure. This makes it a good place to test models like the Aqua Vascular Tower. However, when thinking about implementing similar projects in other tropical megacities such as Jakarta, Dhaka, and Mumbai, several factors need attention:

- Construction costs and material availability might limit use in lower-income areas unless the design is adjusted or simplified.
- Supportive policies for water recycling, vertical greenery, and mixed-use towers are important for making these projects financially viable.
- Cultural views on vertical living and shared spaces vary greatly and should influence the design.

While the design works well in Singapore, its success in other locations will depend on careful adjustments, cost management, and supportive policies.

7.3 Comparison with Existing Vertical Green Projects

Compared to precedent projects like Marina One, Oasia Hotel Downtown, and Bosco Verticale, the Aqua Vascular Tower introduces a key innovation: performance-based biomimicry as the driver of form, not as an aesthetic layer. The performance metrics presented below are compiled from published project documentation and peer-reviewed studies, and are intended to provide comparative context rather than exact equivalence. Bosco Verticale remains one of the most influential precedents for integrating large-scale vegetation into high-rise architecture, demonstrating the ecological and microclimatic benefits of vertical forest typologies [11].

This comparison (Table 2) illustrates how the Aqua Vascular Tower moves beyond current green tower paradigms by embedding environmental functionality directly within architectural form, informed by the operational logic of biological systems. Recent reviews of greenery-covered tall buildings have highlighted their contributions to urban cooling, environmental performance, and occupant well-being, while also identifying maintenance and scalability challenges in high-rise applications [12].

The comparative performance indicators presented in Table 3 are derived from a synthesis of prior studies on vertical greening, facade-based water harvesting, and passive cooling strategies in high-rise buildings.

7.4 Future Research Opportunities

The project opens several paths for academic and technological exploration:

- Computational modeling of vascular performance in architectural skins
- Smart water systems with responsive sensors and capillary-inspired flows
- Data-driven biophilic design metrics to measure psychological benefits
- Integration of urban biodiversity networks into high-rise façades

These opportunities indicate that the Aqua Vascular Tower is not just a design proposal; it serves as a research platform where architecture, environmental science, and health meet.

7.5 Limitations of the Current Study

As a conceptual design that hasn't gone through complete engineering checks, the tower's performance metrics rely on simulated or estimated figures. Future development would need:

- Detailed structural analysis
- Material lifecycle assessments
- Long-term operational feasibility studies
- Feedback from potential users and city authorities

Despite these limitations, the design's clear intent and internal logic give a strong foundation for experimental prototyping or smaller-scale implementation. The Aqua Vascular Tower sits between vision and practicality. It offers a bold, systemic approach to climate-resilient, socially inclusive urban development, while recognizing the challenges of turning biomimicry into physical structures.

Table 2 Comparison with existing vertical green project

Project	Marina One (Singapore)	Oasia Hotel Downtown (Singapore)	Bosco Verticale (Milan)	Aqua Vascular Tower (Proposed)
Primary strategy	Central green heart and sky gardens	Vegetated facade and sky gardens	Vertical forest facade	Biomimetic vascular facade inspired by the human cardiovascular system
Water reuse/ harvesting	Rainwater harvesting supports irrigation of ~37,000 m^2 of landscaped areas	Green façade (~33,000 m^2 planted mesh) enables passive irrigation and rainwater capture	Facade vegetation includes ~800 trees, 4500 shrubs, and 20,000 plants, equivalent to ~5 ha of woodland	Integrated rainwater harvesting and vertical circulation channels designed to support hydroponic irrigation and greywater reuse
Cooling/ microclimate impact	Microclimate cooling of up to ~2 °C through dense vegetation and airflow modulation	1100% green plot ratio enhances shading and reduces solar heat gain, supporting passive cooling	Vegetation shading reduces indoor temperatures by ~2–3 °C and solar radiation by ~49%, lowering cooling demand by ~30%	Combined passive ventilation, evaporative cooling, and façade porosity aimed at reducing heat gain and enhancing airflow at multiple elevations
Biodiversity support	~350 plant species including ~700 trees supporting urban flora and fauna	Biodiversity audit identified ~18 animal species, representing ~50% of surrounding urban species	Supports ~1600 birds and insects annually, fostering high urban biodiversity	Modular hydroponic and vegetated zones intended to support plant diversity and micro-habitats throughout the tower height
Key limitation	Green systems primarily concentrated at podium and atrium; facade remains largely conventional	Performance dependent on plant maturity and intensive maintenance	High structural loads and complex long-term maintenance of mature trees	Conceptual stage; performance metrics require further empirical validation and detailed engineering studies

7.6 *Practical Feasibility and Implementation Pathways*

The practical feasibility of the proposed Aqua Vascular Tower is supported by both existing built precedents and established environmental technologies currently deployed in high-density urban contexts. Existing research on biomimetic vertical

Table 3 Indicative performance comparison between a conventional high-rise and the Aqua Vascular Tower

Parameter	Conventional high-rise	Aqua Vascular Tower (indicative)
Rainwater reuse	Minimal (roof-only systems)	High (facade-integrated harvesting)
Passive cooling contribution	Low	Moderate–high (evapotranspiration + shading)
Green surface ratio	<10%	>35% (vertical hydroponic integration)
Urban biodiversity support	Limited	Moderate (habitat-forming vegetation)
Social interaction spaces	Limited, floor-bound	Multi-level communal terraces

Source Authors' synthesis based on published studies on vertical greening, facade-integrated rainwater harvesting, and passive cooling in high-rise buildings [13, 14, 15]

greening and environmental performance further supports the feasibility of such integrated systems [16, 17]. While the project is conceptual in nature, its key systems facade-integrated greenery, rainwater harvesting, and passive cooling are grounded in proven applications within comparable climatic and urban conditions.

Large-scale vertical greening has already been successfully implemented in Singapore, demonstrating both constructability and operational viability. Notably, the *Tree House* residential development features over **24,000 m² of vertical greenery** integrated into its façade system and achieved a **Guinness World Record for the largest vertical garden**. According to reported post-occupancy data, the project achieved approximately **15–20% energy savings**, attributed primarily to reduced solar heat gain and enhanced microclimatic cooling [18]. This precedent confirms that extensive façade-based planting systems can be constructed, maintained, and perform effectively in tropical high-rise environments similar to the proposed site context.

From a technical standpoint, peer-reviewed research further validates the environmental performance of vertical greenery systems. A comprehensive review by Pérez et al. [19] demonstrates that green façades and vertical vegetation systems can reduce exterior surface temperatures by **up to 15 °C** and lower indoor cooling demand by **20–30%**, depending on plant density, irrigation strategy, and façade orientation. These findings support the thermal and energy performance assumptions applied in the Aqua Vascular Tower's biomimetic façade design.

In the proposed tower, rainwater harvesting is integrated directly into the vascular façade network, allowing gravity-assisted water conveyance to hydroponic cultivation zones and irrigation points. This approach builds upon established rainwater reuse strategies commonly adopted in Singapore's green buildings, while extending their function through vertical redistribution rather than isolated collection. Maintenance strategies for hydroponic systems are informed by existing modular vertical farming practices, emphasizing accessible maintenance nodes, closed-loop nutrient circulation, and phased plant replacement to ensure long-term operational resilience.

It is acknowledged that full implementation would require detailed engineering coordination, life-cycle cost assessment, and long-term performance monitoring. Nevertheless, when evaluated against existing precedents and peer-reviewed environmental performance data, the Aqua Vascular Tower demonstrates a credible and scalable pathway toward integrating biomimetic hydrological systems within high-rise architecture.

Key Takeaways

- The Aqua Vascular Tower advances high-rise design by integrating biomimicry, climate responsiveness, and social infrastructure into a single vertical system.
- While conceptually feasible, full realization requires advances in material systems, smart sensing, and hydroponic management, highlighting a gap between biological inspiration and current construction practices.
- Comparative analysis shows that the project moves beyond existing green towers by making performance-based biomimicry a form-generating strategy, not an applied aesthetic.
- The model is well suited to highly regulated, innovation-driven cities like Singapore, but broader adoption in other tropical megacities depends on cost adaptation, policy support, and cultural context.
- As a research-driven prototype, the tower provides a platform for future interdisciplinary study in environmental simulation, smart water systems, biophilic performance, and vertical biodiversity.
- Acknowledging current technical limitations strengthens the project's credibility while reinforcing its role as a forward-looking framework for climate-resilient urban architecture.

8 Conclusion

The Aqua Vascular Tower reconsiders vertical architecture. It positions the skyscraper as a living, circulatory system within the urban landscape, rather than just a space for various functions. By taking cues from the human cardiovascular system, the tower combines rainwater collection, passive cooling, natural ventilation, and vertical biodiversity into a cohesive design. Estimates of water collection show that vascular façades can act as self-sustaining structures. The hydroponic gardens and vertical public areas contribute to both environmental and social resilience.

Located in Singapore's dense tropical environment, the design uses local weather patterns to create a model that other high-density cities can follow. Its advantage over other vertical green projects lies in merging environmental performance with the overall design, rather than adding greenery as a secondary element. This approach not only lessens its ecological impact but also promotes mental well-being, inclusivity, and biodiversity. It connects well with modern ideas about sustainable urban living and biophilic design.

However, the project also recognizes issues around technological feasibility, cost, and maintenance. It serves as a speculative yet research-based prototype. Future development needs to improve computational fluid dynamics, smart water systems, and flexible construction techniques to bring this vision to life.

In the end, the Aqua Vascular Tower presents a new idea: that future vertical cities can operate as regenerative ecosystems. Architecture can circulate resources and support life just as natural systems do. This concept opens a path toward urban futures that are adaptive to climate change, socially inclusive, and ecologically integrated.

Key Takeaways

- Climate-resilient architecture benefits from principle-based design approaches rooted in biological systems.
- Biomimetic vascular frameworks enable scalable, adaptive, and regenerative high-rise typologies.
- The Aqua Vascular Tower functions as a design framework rather than a singular object, offering transferable strategies for dense urban environments.
- By integrating water, vegetation, airflow, and social space, the project demonstrates how architecture can act as an active agent of climate resilience.

Acknowledgements The author gratefully acknowledges Supervisor Nadia Monzur for her expert guidance and constructive feedback throughout the preparation of this chapter. Appreciation is also extended to the Department of Architecture, Rajshahi University of Engineering & Technology, for their unwavering support. Special thanks are due to Md Imamul Alam Ishat for his valuable assistance with ANSYS analyses, which significantly contributed to the technical rigor of this work. The author also acknowledges the use of AI tools to enhance clarity, structure, and readability, ensuring the presentation of complex concepts is precise and accessible.

References

1. Vashishta A (2024) Vertical farming: an agricultural revolution. Int J Sci Res Eng Trends 10(5)
2. Snider A (2025) Building biomimetically. Boller Rev. https://doi.org/10.18776/tcu/br/8/185
3. bioSEA (2025) Oasia Hotel Downtown: biodiversity and ecosystem service assessment. bioSEA Project Report
4. Gustafson Porter + Bowman, ingenhoven architects (2018) Marina One Singapore: urban sustainability and green heart microclimate performance. Project documentation
5. Arbabzadeh M, Etessam I, Shemirani MM (2020) Passive thermoregulation in vernacular and biomimetic architecture in hot and arid climate. Iran Univ Sci Technol 30:198–211
6. Badarnah L (2017) Form follows environment: biomimetic approaches to building envelope design. Archit Sci Rev 60(1):1–14
7. Benyus JM (1997) Biomimicry: innovation inspired by nature. William Morrow, New York
8. Sowmya C, Anand M, Indu Rani C, Amuthaselvi G, Janaki P (2024) Recent developments and inventive approaches in vertical farming. Front Sustain Food Syst. https://doi.org/10.3389/fsufs.2024.1400787
9. Fewkes A (2012) A review of rainwater harvesting in buildings. Water Sci Technol 66(1):9–17

10. Eisenbarth C, Haase W, Blandini L, Sobek W (2022) Potentials of hydroactive lightweight façades for urban climate resilience. Civ Eng Des 4(1):14–24
11. Studio B (2014) Bosco Verticale: vertical forest. Stefano Boeri Architetti, Milan
12. Greenery-covered tall buildings: a review (2024) Int J Arch Sustain 13:2362.
13. Wong NH, Tan AYK, Tan PY, Chiang K, Wong NC (2010) Thermal evaluation of vertical greenery systems for building walls. Build Environ 45(3):663–672
14. Pérez G, Coma J, Martorell I, Cabeza LF (2014) Vertical greenery systems (VGS) for energy saving in buildings: a review. Renew Sustain Energy Rev 39:139–165
15. Perini K, Rosasco P (2013) Cost–benefit analysis for green façades and living wall systems. Build Environ 70:110–121
16. Gao J (2025) Integrating biomimetic vertical greening systems with plant-cell-inspired design for urban cooling and energy efficiency: a case study in Suzhou. Mol Cell Biomech. https://doi.org/10.62617/mcb1353
17. Kshirsagar MP, Kulkarni S, Meena AK, D'costa DC, Bhagwat A, Siddiqui MIH, Dobrotă D (2025) Biomimicry-based design of underground cold storage facilities: energy efficiency and sustainability. Biomimetics 10(2)
18. BDC Network (2014) Singapore's Tree House vertical gardens break Guinness World Record. Building Design + Construction Network
19. Pérez G, Perini K et al (2017) Green vertical systems for buildings as passive systems for energy savings. Appl Energy 194:247–256

Transforming Industrial Heritage into Climate-Resilient Cultural Spaces: Comparative Insights from Coal Drops Yard (UK) and Argo Factory (Iran)

Laleh Ramezani and Mahdiar Nezam

1 Introduction

Today, cities are facing multiple challenges, such as climate change and rapid development, which have placed the built environment and cultural heritage at increasing risk. In response to these pressures, the adaptive reuse of existing structures has emerged as a vital strategy at the intersection of innovative architecture, heritage conservation, and environmental resilience. English Heritage's Conservation Principles [15] consider adaptive reuse as a change management strategy to sustain significance, keeping places in active use through continual adaptation and the optimal, viable use. In practice, this approach advances sustainability and resilience by retaining existing fabric, encouraging sensitive material reuse, and requiring that climate-mitigation and adaptation measures be proportionately scrutinised.

Adaptive reuse is the process of transforming existing buildings and spaces for new functions while preserving their historical and architectural features. Over time, it has evolved from a practical solution into a conscious, design-led approach for sustainable urban transformation and development [35]. This Approach represents a sustainable alternative to demolition, offering measurable benefits in terms of reduced carbon emissions, material waste, and embodied energy [9].

The conservation of built heritage is directly linked to environmental dimensions and climate change. Given that one-third of global greenhouse gas emissions are attributed to the building sector, the preservation and adaptive reuse of historic structures is emerging as a key strategy for reducing emissions and minimising

L. Ramezani (✉)
School of the Built Environment, Oxford Brookes University, Oxford, UK
e-mail: l.ramezani@brookes.ac.uk

M. Nezam
Manchester School of Architecture, Manchester, UK

A. Cheshmehzangi et al. (eds.), *Design Innovation for Climate-Resilient Cities and Built Environments*, Urban Sustainability, https://doi.org/10.1007/978-981-92-2297-1_6

the carbon footprint associated with demolition [47]. Climate pressures around the world have led contemporary urban development discourse to regard obsolete and unused industrial plants, along with degraded urban spaces, not as liabilities but as resources that can enter a productive cycle of recovery and reuse, providing raw material for contemporary architecture and urban planning. If adaptive reuse projects are designed to meet today's sustainability standards, this approach can offer a practical response to urban management challenges and serve as a lever for climate resilience by extending building lifespans, reducing material and energy demands, and delivering social value through cultural and civic programming [6, 40]. Within the city planning, it has become a creative solution and critical challenge, offering opportunities to align past narratives with contemporary environmental and social needs and the future life of buildings and cities.

This chapter aims to conduct a cross-cultural comparative analysis of two internationally awarded adaptive reuse projects, Coal Drops Yard in London and Argo Factory in Tehran, as a means of evaluating how industrial heritage can be reimagined through design-led innovation to meet both environmental and cultural needs. These two cases, selected from highly distinct socio-political and climatic contexts, provide a robust foundation to examine the intersection of climate resilience, cultural memory, and adaptive architectural practice. The study applies a six-dimensional framework drawn from international literature, and further introduces three emergent values, Dynamic Performance, Structural Maturity, and Creativity, which together inform a holistic understanding of climate-conscious reuse. The central research question guiding this investigation is: How can the adaptive reuse of industrial heritage through innovative design strategies contribute to climate resilience while maintaining cultural continuity in diverse urban contexts? By addressing this question, the chapter contributes to global discourse on sustainable regeneration and offers transferable design insights for heritage-led transformation in climate-vulnerable cities. It conceptualises climate resilience as an integral dimension of urban resilience, in which adaptive reuse enhances environmental performance through reductions in carbon and energy demand, supports social resilience through the continuity of cultural memory and public space, and strengthens economic resilience by enabling regeneration and long-term urban viability. This framing aligns adaptive reuse with contemporary resilience scholarship that emphasises the interdependence of environmental, social, and economic systems in shaping resilient cities [22, 24, 42]. This research draws upon the doctoral thesis of the first author, completed at the University of Tehran, and the academic research of the second author, conducted at the Manchester School of Architecture.

2 Theoretical and Contextual Background

2.1 Adaptive Reuse and Urban Regeneration

A rethinking of the concept of conservation emerged in the 1964 Venice Charter, which highlighted the importance of adaptive reuse as a legitimate act of conservation. The 1970s can be regarded as a pivotal historical moment in which adaptive reuse was established as a creative discipline with philosophical and theoretical foundations [36, 54]. Over time, the focus of conservation expanded from individual monument-like buildings to larger building groups and historic urban fabrics. In this context, adaptive reuse has gained increasing attention as a strategy for maintaining urban vitality and integrity [38].

From the second half of the twentieth century into the twenty-first, adaptive reuse of existing buildings, rather than demolition and replacement, saw significant growth [11]. This shift was particularly driven by changing urban conditions, growing concerns about the "freezing" or musealization of cultural assets, and the challenge of dealing with a large volume of vacant and underutilised buildings in cities [10].

The reuse of old and historic abandoned buildings has increasingly been proposed during urban regeneration and conservation decision-making processes, where it is often seen as a driving force for sustainable development [3, 30, 53].

In parallel with broader shifts in urban regeneration theory, adaptive reuse has increasingly been reframed not only as a conservation strategy, but as an urban regeneration mechanism capable of addressing environmental, social, and economic challenges simultaneously. Contemporary scholarship emphasises reuse as a means of extending building life cycles, reducing demolition-related waste, and limiting the release of embodied carbon associated with new construction [6]. Within regeneration discourse, adaptive reuse is understood as a resource-efficient alternative to replacement-led development, particularly in dense urban contexts where vacant or underutilised building stock represents both a spatial and environmental liability [28, 55].

More recent research situates adaptive reuse within climate mitigation agendas, arguing that retention and transformation of existing structures can significantly outperform new-build solutions when whole-life carbon impacts are considered [18, 47]. Rather than viewing heritage constraints as barriers to sustainability, scholars increasingly recognise reused buildings as contributors to low-carbon transitions, offering opportunities to combine passive environmental performance, material efficiency, and cultural continuity [31]. In this sense, adaptive reuse operates at the intersection of climate responsibility and cultural sustainability, aligning conservation objectives with broader urban resilience and regeneration goals [41, 56].

2.2 *Industrial Heritage and Climate-Resilience Adaptive Reuse*

Following widespread de-industrialisation in the late twentieth century, particularly across Western Europe and North America, large numbers of factories and industrial complexes were abandoned, producing extensive post-industrial landscapes [48]. Initially perceived as economic failures or environmental burdens, these sites increasingly became the focus of cultural preservation and adaptive reuse as cities sought to respond to the economic and social crises associated with industrial decline. As post-industrial urban identities evolved, the architectural and cultural significance of industrial buildings began to be re-evaluated, leading to a gradual shift in urban regeneration strategies from demolition towards reuse [37]. This reorientation positioned industrial heritage as a critical field within adaptive reuse discourse, linking cultural value, urban transformation, and long-term sustainability within contemporary city-making processes [44].

In response to the growing recognition of the cultural and architectural value of abandoned industrial sites, international efforts emerged to formalise their protection, most notably through the establishment of TICCIH in 1978 [43]. Recognised by ICOMOS, TICCIH has shaped key conservation frameworks that position adaptive reuse of industrial heritage as central to sustainable and climate-responsive industrial heritage management [20, 38, 49]. Within this framework, the adaptive reuse of industrial heritage is increasingly understood not only as a strategy for climate mitigation but as a form of cultural sustainability that sustains collective memory, place identity, and everyday urban life while enabling adaptation to environmental change [23, 46].

One of the foundational principles of climate-resilient adaptive reuse is the preservation and optimisation of embodied carbon. Retaining existing structures reduces life-cycle greenhouse gas emissions by avoiding the environmental costs associated with demolition and new construction [6, 15]. Alongside this, enhancing passive environmental performance contributes to greater responsiveness. Many industrial and modern-era buildings possess robust material qualities and spatial configurations reflecting their architectural significance and technological innovation, which can be further strengthened through adaptive reuse interventions that utilise passive design strategies such as thermal mass and natural ventilation to improve comfort and reduce energy demand [14, 37].

Another key principle of climate-resilient adaptive reuse is reversibility and minimal intervention. This approach prioritises preserving architectural character through subtle and respectful changes while ensuring that future adaptations remain possible. Reversible strategies, including dry assembly, detachable systems, and non-invasive retrofits, help maintain material authenticity and allow new elements to remain distinguishable yet compatible with historic fabric [8, 50]. When combined with targeted low-carbon technologies, such interventions support climate mitigation objectives without compromising cultural integrity. Importantly, industrial heritage is thereby understood not as a relic of a carbon-intensive past, but as a site of living memory and identity [47].

To fully embed industrial buildings into the contemporary city, the principle of urban connectivity becomes essential. Buildings should not be treated as isolated monuments; rather, they must remain embedded within the life of the city, contributing to its preservation and vitality. Their position within the urban context and their relationship to the surrounding fabric enhance the dynamic quality of place. Through ongoing and imaginative reuse, buildings can retain meaning and actively participate in contemporary culture. While isolation renders them static relics, integration and urban continuity transform them into living structures with the capacity to evolve over time [21, 44].

Finally, cultural memory and innovation are mutually reinforcing values in climate-resilient adaptive reuse. In this context, adaptive reuse becomes a means of safeguarding collective memory and social value, enabling communities to retain meaningful connections to post-industrial landscapes while embracing sustainable futures [33]. Post-industrial environments are increasingly understood not as remnants of decline, but as culturally productive spaces that support identity, reinterpretation, and community-centred revitalisation [45]. The goal of adaptive and sustainable reuse is therefore to establish a creative dialogue between past and future, while preserving the architectural and cultural characteristics that define built heritage [34]. These six interconnected dimensions, including preserving embodied carbon, enhancing passive performance, enabling reversible and minimal intervention, integrating low-carbon technologies, supporting urban connectivity, and celebrating cultural memory through innovation, together provide a coherent framework for understanding the role of industrial heritage in climate-responsive regeneration. Rather than functioning as abstract principles, they establish an analytical structure through which adaptive reuse projects can be systematically interpreted and evaluated. This framework forms the conceptual basis for the comparative case studies examined in the following section and enables an assessment of how climate mitigation and cultural sustainability are negotiated in practice.

2.3 Cultural and Socio-Political Contexts: UK and Iran

- Kings Cross in London (post-industrial creative regeneration):

Located just north of central London, King's Cross has evolved from a rural core and Roman crossing to one of the city's most important sites for post-industrial regeneration. Historically known as Battle Bridge, the area is steeped in deep layers of mythology, religious tradition, and industrial innovation, including the legendary last stand of Queen Boudica and the founding of one of the oldest Christian sites in Europe. The area was a semi-rural area until the eighteenth century, before the arrival of the Regent's Canal (1820) and the Imperial Gasworks (1824) paved the way for full industrialisation.

By the mid-nineteenth century, King's Cross had become a vital hub for transport, logistics, and industrial production, with coal warehouses, gasworks, and a large railway terminus shaping its economic and urban form. However, the decline of rail transport and industry during the twentieth century left much of the area derelict and underused, causing serious social and urban challenges. Since the early 2000s, a major regeneration scheme, including the redevelopment and adaptive reuse of Coal Drops Yard and the restoration of St Pancras Station, has transformed King's Cross into a leading example of creative reuse. The area today serves as a case study in how industrial heritage, infrastructure investment, and urban culture can be combined to serve social and climate-resilient redevelopment. Today, King's Cross is recognised as one of the UK's most sustainable regeneration projects, integrating energy efficiency, biodiversity, and responsible construction. With over 40% green space, on-site renewable energy, and BREEAM-certified buildings, the development demonstrates a long-term commitment to environmental resilience and climate-conscious urban planning.

In 2023, King's Cross became the first major urban regeneration project in the UK to achieve carbon neutrality. This milestone reflects over two decades of sustainability planning, including reductions in operational emissions, 100% renewable electricity sourcing, and investment in verified carbon offsetting projects aligned with the UN Sustainable Development Goals [25].

- Historic Fabric of Ferdowsi, Tehran (with Argo as Industrial Exception):

The Ferdowsi Square area, located in the heart of Tehran, is one of the urban textures with prominent and important historical and architectural layers. Developed primarily during the late Qajar and early Pahlavi periods, the district features a rich concentration of monumental buildings, cultural institutions, and aristocratic residences, such as the Ministry of Foreign Affairs compound, Bank Melli headquarters, National Garden (Bāgh-e Melli), and several historic schools and palaces.

This zone has traditionally served administrative, diplomatic, and cultural functions, and has never been an industrial quarter. Within this context, the Argo Factory, established in the 1920s, is one of the few industrial structures interwoven into a dominantly civic and ceremonial landscape. Its transformation into a contemporary art centre underscores a broader shift in how Tehran's historic urban cores are being reimagined through large-scale industrial reuse like King's Cross, but via selective cultural interventions that preserve architectural heritage while reshaping urban identity [38].

3 Methodology

This chapter adopts a qualitative, comparative case study approach to explore how the adaptive reuse of industrial heritage through innovative design strategies can contribute to climate resilience while maintaining cultural continuity in diverse urban contexts. This study aims for analytical generalisation rather than statistical inference, using an in-depth comparison of two distinct adaptive reuse projects to extract transferable design principles [57]. The central aim is to critically evaluate how climate-resilient strategies are embedded in the adaptive reuse of industrial heritage through design-led innovation, using a comparative analysis of Coal Drops Yard (UK) and Argo Factory (Iran).

To investigate this question, the study analyses two award-winning adaptive reuse projects, Coal Drops Yard in London and Argo Factory in Tehran, selected through purposive sampling based on four criteria: International recognition and design excellence (e.g. RIBA Award, Aga Khan Award) Proven long-term public use and cultural activation, Location in dense, historically significant urban cores, Clear documentation and availability of architectural, visual, and narrative sources.

3.1 Data Collection

This research draws on a combination of primary and secondary data. Two internationally awarded adaptive reuse projects, Coal Drops Yard and Argo Factory, were selected as case studies. Fieldwork included on-site observations, photographic documentation, and spatial analysis of design interventions and user experience. Secondary sources included architectural drawings, award jury reports, published articles, and media coverage. Where accessible, interviews and public event records provided additional insight into the design process, cultural impact, and long-term performance of each project.

3.2 Analytical Framework

The analysis was guided by a six-dimensional evaluative framework derived from international literature on adaptive reuse, climate-responsive architecture, and heritage conservation [37, 44, 47].

These dimensions include preserving embodied carbon, enhancing passive environmental performance, enabling reversible and minimal intervention, integrating low-carbon technologies, supporting urban connectivity, and celebrating cultural memory. Together, they provide a structured basis for examining how spatial, environmental, and cultural strategies are embedded within adaptive reuse projects.

The analytical synthesis in this chapter was developed through a structured four-step process that guided the interpretation of the case studies. First, case study data were prepared by extracting design, spatial, environmental, and cultural information from primary and secondary sources and organising this material in relation to the six analytical dimensions. Second, a first cycle of coding examined each case dimension by dimension, enabling systematic and comparable interpretation of project-specific strategies. Third, a second cycle of coding identified recurrent patterns that appeared across multiple dimensions and across both cases, shifting the analysis from individual criteria to cross-dimensional relationships. Finally, these recurring patterns were synthesised into three higher-order design values, Dynamic Performance, Structural Maturity, and Creativity. These values extend the six-dimensional framework by capturing integrative qualities that support long-term adaptability, environmental performance, and cultural continuity in climate-resilient adaptive reuse. The analytical workflow underpinning this process is summarised in Fig. 1.

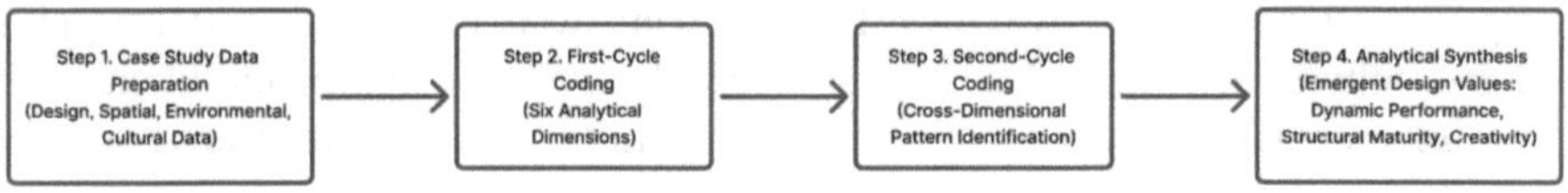

Fig. 1 Analytical workflow for deriving emergent design values

3.3 Interpretation & Research Outcomes

The comparative analysis of the two case studies enabled a cross-case synthesis of how climate-responsive strategies are embedded within adaptive reuse design. By evaluating each project through the six dimensions and the three emergent values, the study identified patterns that are both context-specific and globally relevant. These findings offer transferable lessons for integrating cultural heritage with sustainability goals in diverse urban environments. Ultimately, the research contributes to the development of future-oriented frameworks for adaptive reuse, ones that prioritise not only carbon reduction but also cultural continuity, design flexibility, and long-term urban resilience.

4 Case Studies

4.1 Learn from Coal Drops Yard, London

Coal Drops Yard, as a part of King's Cross, originally built in the 1850s, was a logistical hub for coal arriving in London by rail from the north [26]. The twin viaducts allowed coal wagons to discharge vertically into bays below, creating an early example of multi-level industrial infrastructure [17]. After the decline of coal as a primary fuel source, the buildings were repurposed as warehouses and later housed nightclubs. Despite fire damage in the 1980s, much of the original masonry and cast-iron structure remained intact. The site was later regenerated through an award-winning adaptive reuse project led by Heatherwick Studio in 2018, preserving its historic character while introducing a sculptural new roof [7].

Two long, linear, solid brick and cast iron framed buildings that were once coal sheds as relics of London's industrial heritage, but they were suffering from decay and disuse [25, 26]. Composed of robust brick walls and cast-iron structural columns, the twin sheds represented both industrial heritage and significant embodied carbon. Rather than opting for demolition, the design team at Heatherwick Studio preserved and stabilised the original sheds and reused the historic viaducts and cobbled courtyards, integrating them into a new commercial and public realm while retaining their historic fabric [19, 39]. By repurposing these Victorian elements, the project demonstrates how maximum conservation of existing buildings can be aligned with climate-conscious design and life-cycle carbon reduction goals (Fig. 2).

Fig. 2 Coal Drops Yard: adaptive reuse of Victorian coal sheds into a contemporary urban hub. *Source* Photo taken by the authors

The project has reused many of the original functions and passive spatial features. The retention of the original thick masonry and brick walls and the sloping roof forms, which contribute to thermal mass and natural climate regulation, integrates passive environmental strategies. These features help regulate the building's interior temperature by absorbing heat during the day and releasing it at night, reducing reliance on mechanical systems. The adaptive reuse has retained the natural ventilation paths and shading features inherent in the original structure. In addition, a newly added glazed roof canopy encloses the central courtyard while allowing for ample daylight. Together, these elements enhance occupant comfort and support energy efficiency through a low-impact, climate-friendly design [17, 38]. While passive performance strategies focus on environmental responsiveness through preserving existing structures, respectful transformation and addition can simultaneously be achieved through the creative integration of new architectural elements.

Coal Drops Yard features Heatherwick Studio's signature "kissing roof," a lightweight, sculptural, and innovative addition suspended from above without heavily burdening the historic structure. Suspended walkways and reversible steel insertions demonstrate minimal intrusion, preserving façades and timber trusses while enabling architectural distinction through non-invasive transformation [7, 39].

The addition of new elements and contemporary interventions is an integral and necessary part of the adaptive reuse process. A key factor in the success of such projects, and in their classification as sustainable regeneration initiatives, is the incorporation of low-carbon technologies within the new architectural additions. Coal Drops Yard integrates low-carbon technologies alongside heritage preservation. The project features high-performance glazing, solar-control coatings, and energy-efficient lighting that reduce operational energy use [7]. The redevelopment also includes water reuse systems and optimised heating and cooling through passive design and efficient mechanical systems. Rated under BREEAM criteria, the project demonstrates how environmentally responsible engineering can complement historic architecture, supporting a sustainable, low-carbon urban future [4, 39] (Fig. 3).

Coal Drops Yard plays a critical role in re-stitching the fragmented urban fabric of King's Cross. As Heatherwick describes, the site was previously a "horrible void" behind King's Cross station, derelict, disconnected, and inaccessible. The project transformed this neglected space into a vibrant public destination that opens onto a key pedestrian artery linking the station, Regent's Canal, and Granary Square. By introducing new circulation routes through and around the site, and activating underused arches and courtyards, the design reconnected local communities with this historically industrial zone, integrating it into the broader King's Cross regeneration framework [12] (Fig. 4).

The Coal Drops Yard project preserves the rugged spirit of its industrial past while redefining it for contemporary cultural use. Heatherwick explains that the design sought not to imitate history, but to respond to it, retaining the "bones" of the original structures while inserting new architectural gestures, like the sweeping "kissing roof." Today, the site hosts retail, fashion events, and public gatherings, becoming a venue where London's post-industrial narrative is not just remembered but actively reimagined and lived [12].

4.2 *Learn from Argo Factory, Tehran*

The Argo Factory was originally a beer factory and, like many industrial spaces in city centres, was relocated due to unpleasant odours or pollution, rendering the

Fig. 3 Interior underside of the kissing roof: merging heritage with low-carbon and innovative technologies. *Source* Photo taken by the authors

Fig. 4 Reconnecting the urban fabric through adaptive reuse of Coal Drops Yard: from void to vibrant public space. *Source* Photo taken by the authors

building unused. Despite years of abandonment, its location in Tehran's historic centre and its high spatial potential ultimately attracted the attention of investors, ensuring its continued existence. First built in the 1920s, the structure had remained untouched for four decades until its adaptive reuse into a contemporary art space initiated by the Pejman Foundation in 2020 [32]. This project underwent a compatible reuse process, showcasing various creative stages from which many lessons can be learned. The design preserves the building's industrial materiality while adding spaces that encourage cultural engagement and public access. Creating a spatial experience and user interaction with the building and its role in urban revitalisation were among its strengths. The addition of creative sections contributed to redefining its cultural use, adding to the magnificence of the structure in the present time. Tehran's first independent contemporary art museum is located in a historic former brewery over a century old, which had been unused for many years [2, 51].

The adaptive reuse of the Argo Factory prioritised the conservation of its original brick structure and steel roof framing, preserving the embodied carbon of the early twentieth-century industrial building. Rather than replacing deteriorated elements, the design focused on repair, stabilisation, and selective insertion, minimising material waste and the need for carbon-intensive new construction [5]. This conservation-first approach aligned environmental efficiency with cultural preservation. The project also integrated passive environmental strategies inherent to the original structure. The thick masonry walls provide natural insulation, regulating indoor temperatures by absorbing and releasing heat throughout the day. By exposing existing structural elements and minimising artificial barriers, the design promotes natural ventilation, reducing dependence on mechanical systems. The result is a comfortable indoor environment achieved through low-impact methods.

As noted by the Aga Khan Award jury, the intervention "respected the building's industrial heritage while upgrading its performance for cultural and public use" [1] (Fig. 5).

The Argo Factory's transformation reflects a minimal intervention strategy that carefully preserves its industrial character. Key additions, such as the central sculptural stair and the hovering concrete ceilings ("hats"), were designed to be structurally independent and reversible, allowing the original shell and brickwork to remain largely untouched [52]. Unlike many buildings in downtown Tehran that rely on exposed rooftop HVAC systems, Argo forgoes this conventional approach. Instead, the lattice-like concrete ceilings conceal passive climate strategies while eliminating the need for visible rooftop machinery, creating natural ventilation and a cleaner skyline [51]. These lightweight insertions also allow daylight to filter through while maintaining thermal comfort. As noted by the Aga Khan Award and scholarly assessments, the project blends heritage conservation with sustainable performance, delivering a refined, low-carbon cultural retrofit without resorting to high-energy mechanical systems [1, 27] (Fig. 6).

Located in Tehran's historic urban core, Argo Factory's adaptive reuse has strengthened its connection to the surrounding city by opening formerly enclosed industrial space to public access. The redesign introduced new points of entry, public programs, and urban visibility, positioning the museum as a cultural anchor within a dense and evolving urban fabric [5]. Simultaneously, the project embodies a narrative of memory and transformation. Rather than imitating the past, ASA North inserted a new architectural language that coexists with preserved brick walls and chimney stacks, offering a layered dialogue between heritage and innovation [13]. As Lotfi and Sholeh [29] argue, Argo exemplifies a shift from autocratic to creative reuse in Tehran. Gharaati et al. [16] further highlight its value

Fig. 5 Argo Factory: adaptive reuse of a historic industrial landmark. *Source* Photo taken by Deed Studio

Fig. 6 Sculptural stair reversible, low-impact insertions within Argo Factory's historic brick interior. *Source* Photo taken by Deed Studio

in aligning international adaptive reuse principles with locally embedded cultural continuity, turning industrial memory into an active contributor to contemporary urban identity.

5 Discussion and Implications

5.1 *Beyond Conservation, Towards Climate-Resilience Reuse*

The paired reading of Coal Drops Yard (London) and Argo Factory (Tehran) shows adaptive reuse maturing from conservation-as-preservation to conservation-as-transition, a design-led strategy that couples cultural continuity with climate responsibility. Across different contexts, both projects leverage existing structures to reduce embodied carbon, upgrade passive performance, and re-activate long-dormant urban tissue. The six-dimensional framework clarified how low-impact interventions (minimal/reversible additions, passive microclimates, urban connectivity, and memory-innovation dialectics) together produce resilient places. Yet, the comparative lens also surfaces cross-cutting ideas that sharpen the theory and practice of reuse under climate stress. buildings must perform dynamically across shifting scenarios, exhibit structural maturity by integrating future-ready systems into the inherited fabric, and deploy creativity to translate industrial memory into contemporary cultural life. The remainder of this section synthesises these insights and distils their implications for design, policy, and heritage management.

Table 1 Six-dimensional evaluation of climate-resilient adaptive reuse

Dimension	Coal Drops Yard (London)	Argo Factory (Tehran)
1. Preserve embodied carbon	Retains 1850s brick viaducts, cast-iron frames, and yards; structural stabilisation in place of demolition	Retains brick shell, steel framing, chimney; repair/stabilise rather than replace to conserve stored carbon
2. Enhance passive performance	Thick masonry, large volumes for stack-effect ventilation; glazed canopy admits light while moderating gain/shading	Thermal mass of brick, natural cross-ventilation; lattice/hovering ceilings support daylight and air movement
3. Reversible & minimal intervention	Lightweight "kissing roof," suspended walkways; distinct new–old legibility with limited intrusion	Detached sculptural stair; floating ceilings/additions set off the historic envelope for reversibility
4. Use low-carbon technologies	Efficient lighting/services; components selected to meet high environmental standards with reduced operational energy	Passive-first approach; concealed environmental strategies avoid rooftop HVAC clutter, lowering operational intensity
5. Support urban connectivity	Re-stitches King's Cross; opens new routes linking station, canal, and public squares; activates former voids	Opens a once-closed industrial plot to public programmes; new access points integrate with the historic core
6. Celebrate cultural memory while innovating	Industrial scale/spatial rhythm preserved; bold roof form reframes heritage for contemporary use/events	Brewery identity legible through materials and volume; contemporary art/making reinterprets the site's narrative

5.2 Comparative Analysis (Six Dimensions)

The comparative analysis of international award-winning projects, such as Coal Drops Yard in London and Argo Factory in Tehran, reveals that adaptive reuse has evolved into a climate-responsive design approach that integrates environmental, spatial, and cultural dimensions. Across distinct political, cultural, and planning contexts, both projects demonstrate how existing structures, through innovation, can be reactivated to reduce carbon, support new urban programs, and preserve collective memory. To summarise these observations, Table 1 presents a side-by-side comparison of how each project aligns with the six dimensions of climate-resilient adaptive reuse used in this study.

5.3 Carrying Heritage Forward: Design Principles for Generational Continuity

While the six-dimensional framework provided a practical structure for evaluating climate-resilient adaptive reuse in both case studies, the cross-comparison of Coal Drops Yard and Argo Factory revealed deeper design values that govern the long-term viability and cultural resonance of such projects. These values, Dynamic Performance, Structural Maturity, and Creativity, emerged not only as recurring themes across both contexts but also as essential ingredients for ensuring that reused buildings remain relevant, sustainable, and meaningful into the future.

Dynamic Performance refers to a building's capacity to accommodate multiple functions, shifting usage patterns, and emerging climatic or societal demands. In both projects, adaptable spatial configurations, whether modular galleries in Argo or hybrid retail–public environments in Coal Drops Yard, enable programmatic resilience. This kind of performance does not depend on rigid typologies, but on architectural flexibility embedded in structure, circulation, and environmental control.

Structural Maturity reflects the ability of a reused building to integrate sustainable design strategies with the potential of its inherited physical structure. It includes the preservation of embodied carbon, the use of passive systems, and the capacity to retrofit future energy and environmental technologies without damaging historical integrity. Both case studies demonstrate this principle by retaining original masonry shells and minimising invasive interventions, thereby extending the lifecycle of materials while enhancing functional performance.

Creativity is the connective tissue between heritage and contemporary life. It is what enables architects and conservators to infuse historical structures with new cultural meaning, aesthetic, and vitality. This is visible in Heatherwick's "kissing roof" and ASA North's sculptural stair, both bold additions that reinterpret rather than mimic the past. In doing so, creativity ensures that reused buildings are not only preserved for the sake of memory but actively embraced as platforms for the imagination.

Together, these values affirm that meaningful adaptive reuse is not simply about the past; it is about transferring heritage into the future. Buildings that perform dynamically, mature structurally, and inspire creatively are more likely to endure, be loved, and be preserved by future generations. They will form the backbone of low-carbon, culturally rooted cities in an era where permanence must be redefined through adaptability, innovation, and care (Table 2).

5.4 Implications

The comparative analysis of Coal Drops Yard and Argo Factory, grounded in six dimensions and extended through three cross-cutting design values, demonstrates that adaptive reuse is no longer a static preservation tactic but a vital tool for climate-responsive and culturally embedded urban transformation. The findings offer clear implications for practice. Architects and conservation teams are encouraged to view reuse as a creative opportunity to integrate climate performance, flexibility, and cultural meaning. Policymakers and heritage authorities must move beyond rigid preservation towards controlled freedom, enabling thoughtful, future-ready interventions that respect historical integrity. Urban planners and developers

Table 2 Generational design values across Coal Drops Yard and Argo Factory

Design value	Definition	Coal Drops Yard (UK)	Argo Factory (Iran)
Dynamic performance	Ability to support flexible, multi-use programming and future scenarios	Public–retail adaptability; glazed canopy enables thermal/visual flexibility	Modular gallery/event spaces; passive ventilation for varied occupancy
Structural maturity	Integration of sustainability within the retained structure for long-term performance	Retains viaducts; BREEAM components; efficient, reversible insertions	Retains brick shell; floating ceilings avoid HVAC; long-life passive-first systems
Creativity	Reimagining of industrial heritage through bold and respectful design	"Kissing roof" as a sculptural reinterpretation of industrial rhythm	Sculptural stair and lattice ceilings; blending old brewery with contemporary art venue

are reminded that reused industrial sites can anchor sustainable regeneration by enhancing connectivity, cultural value, and carbon reduction.

Alongside design and planning considerations, the findings underscore the importance of appropriate financial and governance arrangements in enabling climate-resilient adaptive reuse, particularly in contexts characterised by limited resources or heightened climate exposure. The two case studies demonstrate that long-term investment horizons, phased development strategies, and collaborative governance structures involving public authorities, private actors, and cultural institutions are central to the delivery of complex reuse projects. Flexible funding approaches, such as blended public and private investment, cultural endowments, and regeneration-linked value capture mechanisms, can help distribute financial risk while supporting long-term stewardship. At the governance level, adaptive regulatory frameworks that balance conservation objectives with environmental and social priorities are essential for enabling innovation without compromising heritage significance. Collectively, these conditions position adaptive reuse not only as a design-led response, but as an actionable policy instrument capable of aligning climate action, cultural continuity, and economic viability across diverse urban contexts. For researchers and educators, this study proposes a hybrid framework, rooted in six analytical dimensions and enriched by three design values, that can be applied to diverse urban settings. Together, these insights position adaptive reuse as a strategic design approach for building more resilient, low-carbon, and culturally meaningful cities.

6 Conclusion

This chapter has examined how the adaptive reuse of industrial heritage, when guided by design-led innovation, can contribute to climate resilience while maintaining cultural continuity across diverse urban contexts. Through a comparative analysis of Coal Drops Yard in London and Argo Factory in Tehran, the findings directly answer the central research question by demonstrating that adaptive reuse supports climate resilience not as a singular technical intervention, but as a multi-dimensional urban resilience strategy. Across both cases, resilience is achieved through the combined effects of reducing carbon and energy demand, preserving cultural memory and public space, and enabling long-term economic regeneration and urban viability.

The application of the six-dimensional evaluative framework clarifies how these outcomes are produced through concrete design strategies, including the preservation of embodied carbon, enhancement of passive environmental performance, reversible and minimal intervention, integration of low-carbon technologies, support for urban connectivity, and the active celebration of cultural memory. Building on this analytical structure, the three emergent values of Dynamic Performance, Structural Maturity, and Creativity explain how reused industrial buildings sustain adaptive capacity over time by accommodating changing uses, integrating future-ready environmental systems within inherited structures, and translating industrial heritage into contemporary cultural life.

Within this framework, adaptive reuse is positioned as a proactive approach to shaping resilient urban futures rather than a response limited to conservation or preservation alone. By aligning environmental performance, social continuity, and economic longevity within a single design logic, the chapter demonstrates how industrial heritage can function as an active urban resource in cities facing climate uncertainty and accelerated transformation. In this sense, adaptive reuse emerges as a strategic tool for resilience-building, capable of guiding policy, design, and governance decisions while ensuring that cultural meaning is carried forward into the future.

References

1. Aga Khan Trust for Culture (AKTC) (2022) Argo Contemporary Art Museum & Cultural Centre. https://the.akdn/en/how-we-work/our-agencies/aga-khan-trust-culture/akaa/argo-contemporary-art-museum-cultural-centre. Accessed 3 September 2025.
2. Ahmadreza Schricker's Narration of the Conversion of Argo Factory into an Art Museum (2018) http://www.onlineartgallery.ir/?m_id=8&id=12727/روایت-احمدرضا-شریکر-از-تبدیل-کارخانه-آرگو-به-موزه-هنری. Accessed 5 September 2025.
3. Aigwi IE, Egbelakin T, Ingham J, Phipps R (2020) A performance-based framework for prioritising obsolete buildings for adaptive reuse interventions in New Zealand. Sustain Cities Soc 52:101861. https://doi.org/10.1016/j.scs.2019.101861
4. ArchDaily (2021) Adaptive reuse as a strategy for sustainable urban development and regeneration.https://www.archdaily.com/970632/adaptive-reuse-as-a-strategy-for-sustainable-urban-development-and-regeneration. Accessed 10 September 2025.
5. ArchDaily (2022, February 13) Argo Contemporary Art Museum & Cultural Centre/ Ahmadreza Schricker Architecture—North. https://www.archdaily.com/979255/argo-contemporary-art-museum-and-cultural-centre-ahmadreza-schricker-architecture-north-asa-north. Accessed 10 September 2025.

6. Baker H, Moncaster A, Remøy H, Wilkinson S (2021) Retention not demolition: how heritage thinking can inform carbon reduction. J Archit Conserv. https://doi.org/10.1080/13556207.2021.1948239
7. Bateman S, Chambers S, Clark E, Hill R, Lenk P, Tattersall S, Walton J (2019) A Victorian heritage site adapted for 21st-century London. Arup J 2:4–13
8. Bertagni S, Boarin P, Zuppiroli M (2018) The dialogue between structural interventions and sustainability criteria in rating systems for cultural heritage: the experience of GBC Historic Building. Int J Arch Herit, 1–23. https://doi.org/10.1080/15583058.2018.1511001
9. Bullen PA, Love PED (2011) A new future for the past: a model for adaptive reuse decision-making. Built Environ Proj Asset Manag 1(1):32–44. https://doi.org/10.1108/20441241111143768
10. Bloszies C (2012) Old buildings, new designs: architectural transformations. Princeton Architectural Press, New York
11. Douglas J (2006) Building adaptation, 2nd edn. Elsevier, Oxford, UK
12. Dezeen (2018, October 30) Thomas Heatherwick explains design for Coal Drops Yard in London in exclusive Dezeen movie [Video]. https://www.dezeen.com/2018/10/30/movie-thomas-heatherwick-coal-drops-yard-video/. Accessed 18 August 2025.
13. Dezeen (2022, May 8) ASA North transforms Tehran brewery into Argo Factory art museum. https://www.dezeen.com/2022/05/08/argo-factory-museum-tehran-asa-north/. Accessed 18 August 2025.
14. de Jonge W (2017) Sustainable renewal of the everyday Modern. J Archit Conserv 23(1–2):62–105. https://doi.org/10.1080/13556207.2017.1326555
15. English Heritage (2008) Conservation principles, policies and guidance for the sustainable management of the historic environment. English Heritage, London. https://historicengland.org.uk/images-books/publications/conservation-principles-sustainable-management-historic-environment/. Accessed 18 August 2025.
16. Gharaati F, Mahdavinejad M, Nadolny A, Bazazzadeh H (2023) Sustainable assessment of built heritage adaptive reuse practice: Iranian industrial heritage in the light of international charters. Hist Environ: Policy Pract. https://doi.org/10.1080/17567505.2023.2261328
17. Heatherwick Studio (n.d.) Coal Drops Yard. https://heatherwick.com/projects/public-space/coal-drops-yard/. Accessed 16 September 2025.
18. Historic England (2020) Heritage counts: There's no place like old homes—re-use and recycle to reduce carbon. Historic England. https://historicengland.org.uk/research/heritage-counts/2019-carbon-in-built-environment/. Accessed 20 August 2025.
19. Huang Y, Wang F, Hiscock AV, Satyarthi J, Smith H (2024) Including embodied carbon in assessing renovation options for industrial heritage buildings: a review and case studies. Sustainability 17(1):72. https://doi.org/10.3390/su17010072
20. ICOMOS–TICCIH (2011) Principles for the conservation of industrial heritage sites, structures, areas and landscapes. https://www.icomos.org/images/DOCUMENTS/Charters/ga2011_industrialheritage_en.pdf. Accessed 20 August 2025.
21. Jones ZM, Zhang X (2024) Identifying the role of industrial heritage in the European Capital of Culture programme. Built Herit 8(20). https://doi.org/10.1186/s43238-024-00133-4
22. Fatiguso F, De Fino M, Cantatore E, Caponio V (2017) Resilience of historic built environments: inherent qualities and potential strategies. Procedia Eng 180:1024–1033. https://hdl.handle.net/20.500.14243/343987
23. Fusco Girard L, Gravagnuolo A (2018) Circular economy and cultural heritage/landscape regeneration. Circular business, financing and governance models for a competitive Europe. Bioresour Bioprocess 5(1):1–15. https://hdl.handle.net/20.500.14243/343987
24. Meerow S, Newell JP, Stults M (2016) Defining urban resilience: a review. Landsc Urban Plan 147:38–49. https://doi.org/10.1016/j.landurbplan.2015.11.011
25. King's Cross Central Limited Partnership (n.d.-a) About the development. https://www.kingscross.co.uk/about-the-development. Accessed 18 September 2025.

26. King's Cross Central Limited Partnership (n.d.-b) The Coal Drops. https://www.kingscross.co.uk/coal-drops. Accessed 18 September 2025.
27. Khakzand M, Porkar MJ (2022) Development of rejuvenation process in Iranian industrial heritage based on grounded theory. Space Ontol Int J 11(4):17–30
28. Langston C, Wong FKW, Hui ECM, Shen LY (2013) Strategic assessment of building adaptive reuse opportunities in Hong Kong. Build Environ 59:170–181. https://doi.org/10.1016/j.buildenv.2007.10.017
29. Lotfi S, Sholeh M (2020) Adaptive reuse gradient from 'autocratic' to 'creative': a context-based anthology of adaptive reuse experience in Tehran (1970–2020). Int J Arch Herit. https://doi.org/10.1080/15583058.2020.1734686
30. Nasser N (2003) Planning for urban heritage places: reconciling conservation, tourism, and sustainable development. J Plan Lit 17(4):467–479. https://doi.org/10.1177/0885412203251147
31. Normandin K (2021) Climate emergency and our built environment. J Archit Conserv 27(3):147–150. https://doi.org/10.1080/13556207.2021.1992839
32. Pejman Foundation (n.d.) About Argo Factory. https://pejman.foundation/argofactory/about/. Accessed 21 September 2025.
33. Pendlebury J, Townshend T, Gilroy R (2004) The conservation of English cultural built heritage: a force for social inclusion? Int J Herit Stud 10(1):11–31. https://doi.org/10.1080/1352725032000194222
34. Pyburn J (2017) Architectural programming and the adaptation of historic modern era buildings for new uses. J Archit Conserv 23(2–3):95–112. https://doi.org/10.1080/13556207.2017.1312760
35. Plevoets B, Van Cleempoel K (2011) Adaptive reuse as a strategy towards conservation of cultural heritage: a literature review. Trans Built Environ 118:155–164
36. Plevoets B, Van Cleempoel K (2013) Adaptive reuse as an emerging discipline: a historic survey. In: Cairns G (ed) Reinventing architecture and interiors: a socio-political view on building adaptation. Libri Publishing, London, pp 13–32
37. Plevoets B, Van Cleempoel K (2019) Adaptive reuse of the built heritage: concepts and cases of an emerging discipline. Routledge, New York
38. Ramezani L (2024) Developing a comprehensive adaptive reuse process for Iran's contemporary built heritage. Doctoral dissertation, University of Tehran
39. Royal Institute of British Architects (RIBA) (2019) Coal Drops Yard—RIBA London Award Winner 2019. https://www.architecture.com/awards-and-competitions-landing-page/awards/riba-regional-awards/riba-london-award-winners/2019/coal-drops-yard. Accessed 21 September 2025
40. Santamaría Varas M, Mínguez García MC (2022) Architecture, culture, and industry: reflections on the architectural intervention in the CaixaForum Madrid. J Archit Conserv 28(1):59–75
41. Sharifi A, Yamagata Y (2016) Principles and criteria for assessing urban energy resilience: a literature review. Renew Sustain Energy Rev 60:1654–1677. https://doi.org/10.1016/j.rser.2016.03.028
42. Sharifi A (2021) Urban resilience assessment: mapping knowledge structure and trends. Sustainability 13(11):5918. https://doi.org/10.3390/su12155918
43. Smith S (2013) The work of TICCIH. In: Douet J (ed) Industrial heritage re-tooled: the TICCIH guide to industrial heritage conservation. Routledge, pp 9–14
44. Stone S (2019) UnDoing buildings: adaptive reuse and cultural memory. Routledge. https://doi.org/10.4324/9781315397221
45. Storm A (2014) Post-industrial landscape scars. Palgrave Macmillan
46. Soini K, Dessein J (2016) Culture–sustainability relation: towards a conceptual framework. Sustainability 8(2):167. https://doi.org/10.3390/su8020167
47. Tam L, Whitman CJ, Prizeman O (2025) Global climate change and built heritage. Built Heritage 9(17):1–4. https://doi.org/10.1186/s43238-025-00194-z

48. Tempel N (2012) L-industrial landscapes. In: Douet J (ed) Industrial heritage re-tooled: the TICCIH guide to industrial heritage conservation. Carnegie Publishing Ltd, pp 220–229
49. TICCIH (2003) The Nizhny Tagil Charter for the Industrial Heritage. The International Committee for the Conservation of the Industrial Heritage. https://ticcih.org/about/charter/. Accessed 27 September 2025.
50. Vafaie F, Remøy H, Gruis V (2023) Adaptive reuse of heritage buildings: a systematic literature review of success factors. Habitat Int 142:102926. https://doi.org/10.1016/j.habitatint.2023.102926
51. Volner I (2021) Argo Factory: Tehran's first contemporary art museum since the revolution. PIN–UP Magazine. https://archive.pinupmagazine.org/articles/argo-museum-hope-factory-pejman-tehran-ahmadreza-schricker. Accessed 27 September 2025.
52. Wallpaper* (2022, July 1) Tehran's Argo Factory art museum and cultural centre opens. https://www.wallpaper.com/architecture/argo-factory-contemporary-art-museum-cultural-centre-ahmadreza-schricker-architecture-north-tehran-iran. Accessed 27 September 2025.
53. Worthing D, Bond S (2016) Managing built heritage: the role of cultural values. Wiley Blackwell
54. Wong L (2017) Adaptive reuse: extending the lives of buildings. Birkhäuser
55. Wilkinson SJ, Remøy H (eds) (2018). Building urban resilience through change of use. Wiley Blackwell
56. Wilkinson S, Remøy H, Langston C (2014) Sustainable building adaptation: innovations in decision-making. Wiley Blackwell
57. Yin RK (2018) Case study research and applications: design and methods, 6th edn. Sage.

Housing Futures: Comparative Strategies for Climate-Resilient Housing in the Global North and South

Parsa Ahmadi Dehrashid and Hossein Mansourian

1 Introduction

Housing constitutes not only a fundamental human need but also a cornerstone of sustainable urban development. Yet in the context of accelerating climate change, housing has emerged as one of the most vulnerable domains within the built environment [1]. Extreme weather events, rising temperatures, and shifting precipitation patterns exert unprecedented pressures on housing quality and long-term sustainability [234]. The significance of this challenge is heightened by housing's direct interconnections with public health, livelihoods, infrastructure, and social stability [567].

Despite the urgency of the challenge, the integration of climate-resilience considerations into housing systems in the Global South remains limited. A large proportion of existing housing stock is inadequately designed, constructed, or sited to withstand emerging climatic conditions [8910]. Prevailing policy frameworks are often derived from models of the Global North; these models are grounded in advanced technologies and large-scale investments, and their implementation proves challenging in resource-constrained contexts [11]. Consequently, millions of households in the Global South remain caught in a cycle of vulnerability, in which each new climatic shock further exacerbates their structural fragility [121314]. This raises the central question of the study: which strategies can foster housing systems in developing countries that are simultaneously resilient to climate change and responsive to local contexts?

In recent years, a growing body of literature has examined the intersections between climate change and housing. Several review studies have attempted to

P. Ahmadi Dehrashid · H. Mansourian (✉)
Department of Human Geography and Planning, University of Tehran, Tehran, Iran
e-mail: h.mansourian59@ut.ac.ir

P. Ahmadi Dehrashid
e-mail: ahmadi.parsa@ut.ac.ir

A. Cheshmehzangi et al. (eds.), *Design Innovation for Climate-Resilient Cities and Built Environments*, Urban Sustainability, https://doi.org/10.1007/978-981-92-2297-1_7

develop comprehensive frameworks for conceptualizing housing resilience. For instance, [15], drawing on an extensive review, introduced various models of resilient and sustainable housing, highlighting shared elements such as secure land use, durable construction materials, and alternative sources of water and energy. Similarly, [2], in a systematic review of reviews, examined the nexus of homelessness, housing insecurity, and climate change, showing that low-income and vulnerable populations disproportionately shoulder the impacts of climatic shocks. In addition, a wide range of studies has concentrated on particular socio-spatial contexts [16], adopting a climate justice perspective, demonstrates that low-income households, older adults, and marginalized populations are most severely affected by housing-related impacts of climate change, while simultaneously having the least access to adaptive resources [17], examining Lagos as a representative African megacity, underscored the multidimensional vulnerabilities of housing and the pivotal role of urban governance in shaping resilience. Along similar lines, [18] emphasized the potential of indigenous climatic design embedded in Europe's historic urban fabrics, whereas [19] analyzed the specific vulnerability of social housing to summer overheating. Likewise, [20], adopting a critical lens, cautioned that without deeper theorization, the notion of "resilience" in housing risks being reduced to a mere policy catchphrase. More recent studies, such as [21] in Kenya, underscore the importance of local perceptions and community priorities, reinforcing the need to design context-specific responses. Comparable evidence from Pakistan [22] similarly illustrates that blending indigenous knowledge with modern innovation can yield more effective solutions across diverse climatic contexts.

Despite the expansion of this body of scholarship, several critical gaps remain unaddressed. First, most existing reviews are either confined to the contexts of developed countries or focus exclusively on a single dimension, such as the technical, social, or policy-related aspects, without clarifying the interconnections among these layers. Second, although case studies have been undertaken in developing countries, their findings have not yet been consolidated into a systematic and comparative framework, thereby restricting both their generalizability and their utility for policy design. Taken together, these limitations underscore the need for a comparative and analytical review capable of bridging Global North and Global South perspectives, identifying key commonalities, and simultaneously offering context-sensitive responses to the resource constraints of climate-vulnerable settings. This study is explicitly designed to address this gap by advancing an operational framework.

This study employs qualitative content analysis of academic literature, design guidelines, and national policy documents in order to address this gap. By juxtaposing cases from the Global North and Global South, the analysis distills strategic patterns and translates them into a comparative and operational framework. The proposed framework emphasizes adaptability, scalability, and grounding in practical experience, thereby offering direct applicability for policymakers, urban planners, and housing designers. Accordingly, this study constitutes an effort to bridge theoretical debates with the lived realities of at-risk communities, while offering pathways that are both scientifically grounded and operationally feasible for shaping the future of housing in developing countries.

2 Methodology

To ensure transparency, coherence, and reproducibility in data collection and analysis, this study adopted the PRISMA framework (Fig. 1). Widely regarded as one of the most robust protocols for systematic reviews, PRISMA facilitates the precise documentation of search strategies, screening procedures, and resource selection [23, 24]. The initial literature search was conducted in the Web of Science (WoS) database, given its extensive coverage of peer-reviewed articles across disciplines relevant to housing, urban planning, and climate change. For analytical consistency, only sources published in English were included. The search process was completed on 11 May 2025. To identify relevant articles, a comprehensive and purposively designed search string was constructed, incorporating widely used keywords related to "housing," "resilience," and "climate change." The search string was defined as follows:

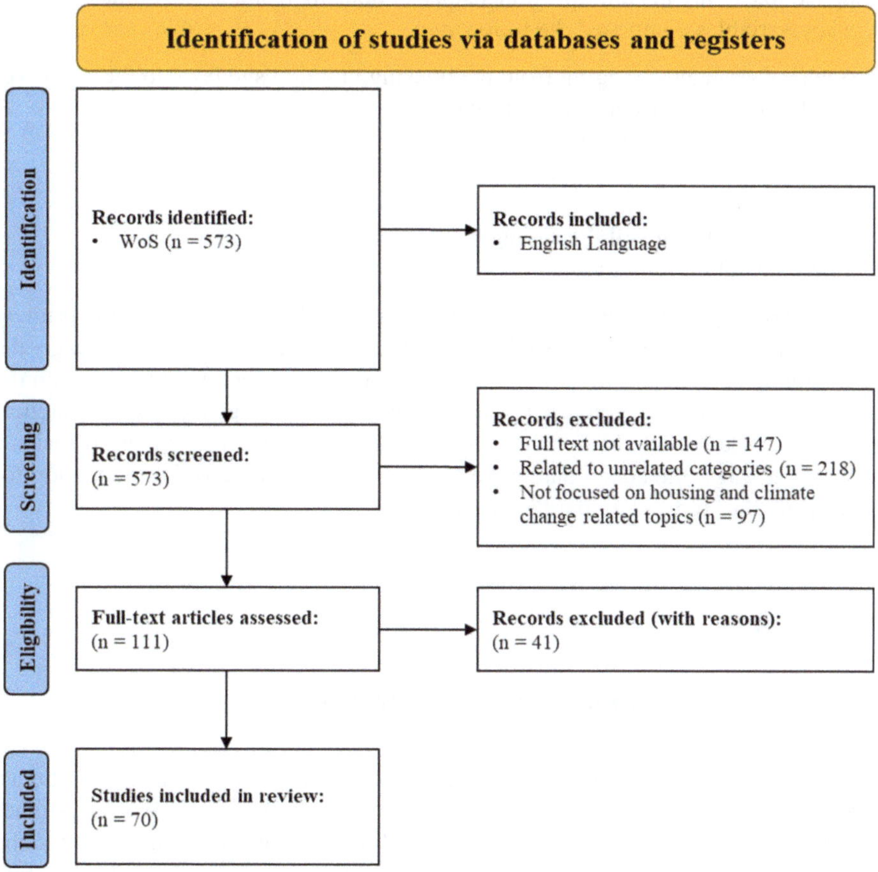

Fig. 1 PRISMA flow diagram of the study selection process

The search in the WoS database was restricted to titles, abstracts, and keywords, yielding 573 initial records. In the first step, records lacking full-text access were excluded (N = 147). At the same stage, studies focusing on unrelated domains, such as water and energy resources, were also excluded (N = 218). During the screening stage, the titles, abstracts, and keywords of the remaining records were reviewed, and articles falling outside the scope of housing and planning (e.g., studies on residents' psychological resilience) were excluded (N = 97). In the subsequent eligibility stage, the full texts of the articles were assessed, and those not aligned with the thematic scope of the study were excluded (N = 41). Ultimately, 70 articles that met the inclusion criteria were retained for the final analysis.

To address the research questions, the study employed Directed Qualitative Content Analysis. This approach facilitates the systematic extraction of concepts, themes, and patterns from scientific texts, guided by initial theoretical frameworks [25, 26]. For data analysis, a structured coding protocol was developed and implemented in Microsoft Excel. The protocol encompassed a set of core variables designed to capture the multiple dimensions of housing resilience to climate change. The key variables comprised the following:

- Scale of strategies: ranging from the housing unit and site level to the neighborhood, city, and broader territorial scales;
- Climate-related housing impacts;
- Strategies for housing resilience;
- Level of intervention;
- Anticipated benefits of the strategy;
- Implementation limitations.

The coding process was conducted in two stages. In the first stage, the primary researcher independently reviewed and coded all sources. In the second stage, the co-author systematically reviewed all extracted codes and compared them with the dataset generated by the primary researcher. In cases of divergence, both researchers convened to deliberate, and consensus was reached on the final coding decisions. In the final stage, the coded data were consolidated and used as the foundation for developing the study's comparative framework.

3 Result

3.1 Impacts of Climate Change on Housing

Evidence from both the Global North and Global South demonstrates that climate change exerts profound, multi-scalar impacts on housing. In the Global North, studies highlight heightened risks of flooding, sea-level rise, severe storms, urban heat island effects, and land subsidence, with associated consequences such as construction delays, material degradation, declining insurability of homes, and rising maintenance costs [3, 27, 28]. For example, Hurricane Sandy in 2012 (Fig. 2) was among the most costly climate-related disasters in the United States, causing widespread damage to housing and urban infrastructure. This case illustrates that even in developed countries with substantial financial and institutional capacities, the housing sector remains highly vulnerable to extreme climatic events (Box 1). Nevertheless, post-disaster responses in the United States, including insurance mechanisms, rapid reconstruction, and investment in resilient infrastructure, demonstrate a significantly higher adaptive capacity than that observed in the Global South. Rising summer temperatures contribute to overheating in buildings, escalating demand for cooling, and increasing carbon emissions, while flood and storm damage jeopardize the habitability and economic value of coastal and low-lying settlements [29, 30]. In the Global South, research consistently shows that informal and low-income housing is disproportionately exposed to risk. Evidence indicates that recurrent flooding in cities such as Lagos, Dhaka, and Accra, which is exacerbated by inadequate drainage and unplanned settlements, causes waterlogging, structural damage, and resident displacement, while extreme heat undermines the livability of affordable housing with insufficient design provisions [31]. In certain contexts, vernacular architecture has demonstrated greater resilience to climatic variability [32, 33]. However, rapid urbanization has eroded these adaptive capacities, heightening vulnerability to heat stress, flooding, and climate-sensitive diseases [21, 34]. Recent post-disaster assessments in Malawi provide concrete, on-the-ground evidence that extreme climate events directly translate into severe housing vulnerability, particularly in low-income and informally built urban areas (Box 2, Fig. 3).

Box 1. The Impacts of Climate Change on Housing in the Global North
Climate change has emerged as an unprecedented threat to housing security in the United States. In 2021, natural disasters affected approximately 10 percent of the nation's housing stock, causing nearly $60 billion in damages and displacing more than 570,000 people. Coastal communities, which are home to more than 125 million Americans, are particularly vulnerable to sea-level rise, a phenomenon projected to force over 13 million people to relocate by 2100. Yet the impacts of climate change are not confined to coastal areas. Flooding, wildfires, and climate-induced migration constitute additional and intensifying risks. At the same time, the housing market and the broader financial system, including insurance, mortgage lending, and public budgets, are under severe strain. Despite recent policy initiatives such as the 2022 Inflation Reduction Act, the shortage of approximately five million housing units and the ongoing affordability crisis, compounded by the legacy of historical discrimination such as redlining, have left low-income households and racial minorities disproportionately exposed to climate-related displacement and homelessness.

Fig. 2 Destroyed beach house in Far Rockaway, New York, after Hurricane Sandy (2012). Reproduced from New [35]

Fig. 3 Flood- and landslide-damaged housing following Tropical Cyclone Freddy in Malawi (2023). Reproduced from [36]

Box 2. Housing Vulnerability Revealed Through Post-Disaster Needs Assessment
The Post-Disaster Needs Assessment (PDNA) conducted following Tropical Cyclone Freddy in Malawi [36] *provides clear, on-the-ground evidence of housing vulnerability to climate-related hazards in Sub-Saharan Africa. The cyclone triggered intense flooding and landslides, resulting in widespread destruction of residential buildings, particularly in densely populated and informally built areas.*

The assessment identifies housing damage and loss as one of the most critical impacts of the disaster, directly contributing to large-scale displacement and prolonged shelter needs. By systematically documenting physical damage, recovery costs, and reconstruction priorities, the PDNA illustrates how extreme climate events translate into acute housing vulnerability, underscoring the central role of resilient housing in disaster risk reduction and climate adaptation strategies.

Both regions face common threats from flooding, overheating, and energy stress; however, the intensity and nature of these impacts differ markedly. In the Global North, wealthier housing systems primarily confront financial risks, such as escalating insurance costs, declining property values, and displacement associated with uninsurability, alongside technological pressures to expand cooling capacity and undertake energy-intensive reconstruction [37]. By contrast, in the Global South the consequences are more existential, involving recurrent displacement, structural collapse, and elevated mortality driven by weak infrastructure, high residential density, and the absence of climate-adaptive design [38]. For example, the 2022 monsoon floods in Pakistan (Fig. 4) affected more than 30 million people, resulting in extensive housing destruction, mass displacement, and over one thousand fatalities. This disaster illustrates that in settings with limited institutional capacity and resources, housing vulnerability rapidly escalates into a humanitarian crisis while simultaneously constraining prospects for sustainable reconstruction (Box 3). Interestingly, vernacular architecture has in many instances demonstrated greater resilience to climatic variability than modern concrete or steel housing, suggesting that processes of "modernization" have at times exacerbated vulnerability [39, 40]. Moreover, both regions face rising demand for cooling, in the Global North this challenge is framed largely in terms of cost and greenhouse gas emissions, whereas in the Global South it directly threatens thermal comfort and household affordability. The literature further indicates that studies in the Global North concentrate predominantly on physical and economic dimensions [27, 41, 42], while research in the Global South emphasizes social vulnerability [14, 43, 44]. However, comprehensive analyses that integrate technological, cultural, and governance dimensions remain scarce [15, 17, 45].

Fig. 4 Flood damage in Pakistan (2022). Reproduced from [46]—Kamran Khan/iStockphoto

Box 3. The Impacts of Climate Change on Housing in the Global South
The 2022 floods in Pakistan represented one of the most severe climate-related disasters in the contemporary Global South. Unprecedented monsoon rains affected nearly 30 million people and resulted in the loss of more than one thousand lives. The extensive destruction of housing and the large-scale displacement of communities demonstrated that even dwellings in both rural and urban areas exhibited limited resilience to such extreme events. Response efforts primarily consisted of emergency relief, temporary shelters, and the distribution of food and water. However, the lack of resilient infrastructure and limited financial resources severely constrained prospects for sustainable reconstruction. This case underscores how developing countries, constrained by limited resources and institutional capacity, are disproportionately trapped in a cycle of housing vulnerability to climate change.

Taken together, the evidence demonstrates that climate change is reshaping the housing landscape in the Global North and Global South in fundamentally different ways. These divergences reflect structural disparities in infrastructure, governance, and socio-economic resilience. In the Global North, the consequences underscore the necessity of long-term adaptation investments, such as constructing dams and protective barriers, promoting resilient building practices, undertaking energy-intensive reconstruction, and reforming insurance and property markets to mitigate displacement. In the Global South, interventions must prioritize informal and low-income settlements, encompassing improved drainage systems, passive cooling design, and the adaptation of vernacular architectural principles, alongside measures to reduce entrenched structural inequalities. Although both regions confront rising energy demand and urban heat island effects, the primary challenge in the Global North lies in decarbonizing adaptation processes, whereas in the Global South it centers on ensuring affordable and sustainable access to cooling. Policymakers must balance housing renewal and modernization with the preservation of climate-adaptive practices, recognizing that the integration of social justice and cultural knowledge is essential for the success of adaptation strategies.

3.2 Housing Resilience Strategies in Response to Climate Change

A substantial body of research has proposed strategies for enhancing housing resilience to climate change in the Global North. [27] underscore the importance of avoiding construction in high-risk areas and implementing flood-resistant design, whereas [41] emphasize "climate-resilient urban design," incorporating green spaces, modular water storage, and elevated flooring systems. [42] highlight reconstruction technologies, including high-performance insulation, solar systems, and HVAC optimization, while [30, 47] point to green roofs, vegetated façades, and water-retention systems as effective responses to heat stress and flooding. For instance, the Garden House project in Beverly Hills, California (Fig. 5), exemplifies the integration of nature into architectural design and is widely recognized as a model for climate-resilient housing development (Box 4). Subsequent research has expanded this field: [48, 49] demonstrate the advantages of permeable surfaces, urban vegetation, and evaporative cooling, show that passive design strategies, including courtyards, night ventilation, and shading devices, can deliver up to 88% of annual thermal comfort without reliance on air conditioning. Institutional and governance interventions are equally critical. [28, 50, 51] emphasize policy reforms, insurance mechanisms, and financial instruments as essential components of resilience. More recent scholarship, including Adeyemi et al. [52] and [53], highlights energy-efficient retrofitting, flood-resilient construction, and the integration of housing within land-use planning, whereas [54, 55] stress the importance of legal frameworks, updated flood maps, and infrastructure upgrading.

Fig. 5 Garden House, Beverly Hills, California (2020). Reproduced from Green Roofs for Healthy Cities (2014)

Box 4. An Example of Using Vertical Gardens to Enhance Climate Resilience
The Garden House project in Beverly Hills stands as a prominent example of integrating contemporary architecture with biophilic principles and green design. This mixed-use residential complex features the largest green wall in the United States, spanning 6,700 square feet, incorporating 28 plant species, and comprising more than 40,000 individual plants. Its hydroponic system, which circulates water continuously through layered panels, creates a vertical ecosystem that not only provides environmental functions but also enhances the visual and psychological quality of the space. This project illustrates how, within the Global North context, design innovation can proactively enhance housing resilience and urban quality of life before climate crises occur.

Conversely, the literature documents a wide spectrum of strategies aimed at enhancing housing resilience to climate change in the Global South. Structural and technological interventions encompass passive design strategies, such as insulated roofs, shading devices, wind catchers, and high thermal-mass walls, which mitigate overheating and reduce energy demand [31, 52, 56]. Policy and governance approaches in these contexts emphasize integrating climate risks into land-use planning, strengthening building regulations, and incorporating informal settlements into formal urban development frameworks [17, 57]. Community-based strategies, including participatory flood mapping, improvements to local drainage systems, and resident-led upgrading of settlements, have been shown to strengthen preparedness

Fig. 6 Upgrading of informal housing under the Favela-Bairro programme, Rio de Janeiro, Brazil. Reproduced from: Tea After Twelve ("Brazil Without Favelas") [61]

and adaptive capacity [38, 58]. At the financial and economic level, pro-poor mechanisms, such as microfinance schemes and housing subsidies, have expanded access to more resilient housing options [17]. An illustrative example of how these structural, governance, community-based, and financial strategies can be combined in practice is provided in Box 5, drawing on empirical evidence from Latin America (Fig. 6). Case studies demonstrate that these interventions improve disaster preparedness, reduce flood-related damages, enhance thermal comfort, and foster social empowerment [17, 59, 60].

Box 5. An example of integrated housing resilience strategies in the Global South
The Favela-Bairro program in Rio de Janeiro represents a well-documented example of integrated housing resilience strategies in the Global South. Initiated in the 1990s and supported by municipal authorities and international development institutions, the program aimed to upgrade informal settlements by combining physical, social, and institutional interventions rather than pursuing large-scale displacement.

At the structural and technological level, housing improvements incorporated low-cost design measures, including improved roofing, ventilation, and shading, to enhance thermal comfort and reduce exposure to heat and flooding. At the policy and governance level, the program focused on integrating informal settlements into the formal urban fabric through land-use regularization, infrastructure provision, and alignment with municipal planning frameworks.

Community participation constituted a central component of the intervention, with residents actively involved in the identification of priorities, neighborhood upgrading processes, and maintenance strategies. From a financial perspective, the program relied on public funding complemented by international development loans, enabling incremental housing improvements and expanded access to basic services for low-income households.

Empirical evaluations indicate that Favela-Bairro contributed to improved housing quality, reduced environmental vulnerability, enhanced access to urban services, and strengthened social cohesion. As such, the program exemplifies how a combination of passive design strategies, inclusive governance, community engagement, and pro-poor financing mechanisms can enhance housing resilience in climate-vulnerable contexts across Latin America.

3.3 Benefits of Housing Resilience Strategies

Empirical evidence from numerous studies demonstrates that strategies adopted in the Global North to enhance housing resilience to climate change generate a wide range of benefits. [27] demonstrated that avoiding construction on flood-prone land reduces both the likelihood of project rejection and reputational risks, while off-site construction provides greater control over timeframes and costs. [42] reported substantial reductions in energy consumption and greenhouse gas emissions in warm climates, which were further reinforced by the integration of solar panels. These findings have also been validated in practice. For example, the installation of solar panels in the Stuytown and Peter Cooper Village housing complexes in New York, with a capacity of 3.6 MW, tripled Manhattan's solar electricity generation and supplied the energy needs of more than 27,000 residents (Box 6; Fig. 7). [47] highlighted that resilience strategies not only reduce flood and heat risks but also strengthen social housing organizations and deliver measurable socio-environmental benefits. [30, 62] demonstrated that reflective coatings and water-storage systems can reduce surface and indoor temperatures, increase property market value, and enhance thermal comfort. Other studies emphasize the effectiveness of green infrastructure [48, 50], passive cooling systems [636465], financial incentives [51], and legal frameworks [66] in reducing climate risks, strengthening housing resilience, and generating socio-economic co-benefits.

Fig. 7 Solar installation at Stuytown and Peter Cooper Village, New York (2020). Reproduced from: Onyx Renewable Partners

Box 6. An Example of Solar System Installations in Urban Multifamily Housing
The solar panel installation at the Stuytown and Peter Cooper Village residential complexes in New York, with a total capacity of 3.6 megawatts, represents the largest private solar deployment in the U.S. multifamily housing sector. Implemented across 62 residential buildings, the project tripled solar energy production in Manhattan and now supplies electricity to approximately 27,000 residents.

A growing number of studies have examined strategies for strengthening housing resilience in the Global South. Evidence indicates that these strategies generate a wide spectrum of benefits in the face of climate-related threats. For example, [59] emphasized the importance of strengthening preparedness, reducing vulnerability, and improving local governance. In Lagos, [17] demonstrated that a combination of formal and informal initiatives reduced flood-related damage and enhanced housing security. From an energy efficiency and thermal comfort perspective, [31, 56, 67, 68] demonstrated that measures such as shifting activity away from peak heat hours, increasing thermal mass, employing night ventilation, and reducing cooling energy consumption by over 50% can deliver substantial improvements. More recent studies, emphasize reducing flood impacts, strengthening social cohesion, and facilitating decision-making in post-disaster housing reconstruction [697071]. Studies from Iran further highlight the role of indigenous strategies, such as central courtyards, wind catchers, and traditional architecture, in reducing cooling demand, improving indoor air quality, and reinforcing cultural identity [727374]. For instance, wind catchers in desert cities such as Yazd (Fig. 8) have for centuries provided natural cooling and

Fig. 8 Windcatcher in Yazd, Iran. Reproduced from Green [76]

ventilation. These structures, capable of reducing indoor temperatures by 8–12 °C, are now regarded as valuable precedents for sustainable architecture and natural ventilation in contemporary design (Box 7). Finally, recent studies, point to benefits such as improved health and well-being, enhanced structural safety, and reduced maintenance costs [21, 75].

Box 7. An Example of Vernacular Architectural Strategies for Hot-Arid Climates
Windcatchers in traditional Iranian architecture, particularly in the city of Yazd, are exemplary instances of climate-responsive design that provided sustainable cooling solutions centuries before the advent of modern air conditioning. These vertical towers captured wind at higher elevations and directed it into interior spaces, reducing ambient temperatures by as much as 8 to 12 degrees Celsius. In some cases, airflow was channeled across subterranean qanat water systems, generating additional cooling effects. Even on windless days, the chimney effect induced by solar heating facilitated natural ventilation. Windcatchers are recognized not only as environmentally adaptive technologies but also as cultural symbols of Yazd, a city inscribed on the UNESCO World Heritage List. Today, the principles of windcatcher design continue to inspire sustainable architecture and natural ventilation strategies in contemporary projects.

3.4 Limitations of Housing Resilience Strategies

Findings from multiple studies reveal recurring limitations in strategies designed to enhance housing resilience to climate change in the Global North. [41] and [47] highlight institutional complexity and fragmented responsibilities as major obstacles to the coordination of adaptation measures. [42] and [18] demonstrate that retrofit measures in temperate climates tend to be less cost-effective and may prove inadequate under conditions of extreme heat. [27] emphasize the speculative character of cost–benefit analyses, stemming from uncertainties in climate projections and the misalignment between costs borne by developers and benefits realized by future homeowners or insurers. [62, 77, 78] further point to weak policy integration and regulatory inconsistencies as barriers to effective implementation. High upfront costs and financial uncertainties [30, 50, 79], together with social and behavioral resistance to adaptive technologies [80, 81], are consistently identified as key obstacles. In addition, several studies stress the limited transferability of results across climatic, cultural, and urban contexts [18, 82], as well as persistent knowledge gaps and an overreliance on simulation models rather than empirical field evidence [83].

Conversely, numerous studies highlight significant limitations in housing resilience strategies across the Global South when confronted with climate change. [17] identify weak coordination across governance levels, insecure land tenure, and mismatches between technical interventions and local socio-economic conditions as major barriers that undermine long-term investment in adaptation. High upfront costs associated with retrofitting and passive design strategies constitute major obstacles for low-income households [52, 56, 68]. The erosion of traditional knowledge, coupled with conflicts with modern urban regulations, has further constrained the application of indigenous techniques [84]. Fragmented governance arrangements and short-term financing cycles have hindered institutional learning and constrained the scaling-up of resilience initiatives [59]. In addition, uncertainties in climate data, household behavior, and the absence of standardized resilience indicators undermine confidence in proposed adaptation measures [85]. Recent studies further identify weak law enforcement, risks of land commodification, and barriers to community participation as critical challenges to the equitable implementation of resilience strategies [21, 22].

Taken together, the findings of this section highlight the following key insights:

- Climate change affects housing systems in the Global North and Global South in fundamentally different but interconnected ways, shaped by structural, economic, and institutional conditions.
- Housing resilience strategies in the Global North emphasize technological solutions, financial instruments, and regulatory frameworks, while those in the Global South rely more heavily on low-cost innovations, vernacular design, and community-based approaches.
- Both contexts demonstrate measurable benefits from resilience strategies, but each also faces distinct limitations related to cost, scalability, governance, and social equity.

- These findings underscore the need for integrative approaches that bridge technological capacity and social adaptability in addressing housing resilience.

4 Hybrid Strategies for Climate-Resilient Housing

Addressing climate change in the housing sector reflects not only geographical and climatic differences but also deeper structural, institutional, and social divides between the Global North and Global South. In the developed contexts of the Global North, resilience strategies are primarily anchored in advanced technologies, large-scale capital investments, and complex governance systems [66]. Measures such as energy-efficient retrofitting, flood-resilient design, green infrastructure, and sophisticated financial and insurance instruments have proven effective in managing specific climate-related risks [27, 28, 41]. However, these models frequently encounter limitations, such as high upfront costs, dependence on precise climate data, and institutional fragmentation, which constrain their scalability and transferability [37, 42, 79].

By contrast, the Global South grapples with a very different set of realities: widespread informal settlements, fragile infrastructure, constrained financial resources, and fragmented governance. As a result, adaptation strategies there often rely on low-cost innovations, vernacular architecture, community participation, and microfinance mechanisms [17, 22]. Although these approaches have often proven successful in terms of affordability and social participation, their effectiveness is frequently curtailed by limited institutional support, conflicts with formal regulations, and the scarcity of reliable data.

The evidence indicates that neither pathway on its own is sufficient to address the complex housing challenges of the climate change era. Advanced technologies and large-scale investments in the Global North, when decoupled from participatory and social mechanisms, often reproduce existing inequalities. For instance, [86] warn that an exclusive reliance on engineering solutions risks reinforcing socio-economic divides. By contrast, low-cost, community-based innovations in the Global South, lacking sustained institutional and legal support, remain highly vulnerable to erosion or failure.

In this regard, [22] underscore the importance of community-based design solutions, active citizen participation, and enabling policy frameworks in fostering resilience. Accordingly, this study advances a comparative framework premised on the strategic and creative integration of these two pathways: (1) combining the technological and institutional capacities of the Global North with the social flexibility; and (2) indigenous knowledge of the Global South. Within this framework, housing is understood not merely as a physical structure but as a socio-cultural institution that must adapt to climatic pressures while simultaneously addressing the imperatives of social and economic justice.

Within this framework, integrative strategies are proposed not only at the conceptual or policy level but also as operational measures that can be practically implemented. Table 1 and Fig. 15 illustrate how the experiences and capacities of Global

North and Global South contexts can be juxtaposed in ways that neither substitute one for the other nor devolve into one-sided imitation, but rather generate "hybrid" and adaptive models. This perspective offers a practical framework for policymakers and planners to move beyond the binary of two worlds and advance toward collaboration and mutual learning.

At the physical level (Fig. 9), experiences in the Global North have largely centered on technological advances such as high-performance insulation, the integration of renewable energy systems, and climate-resilient structural design. While technically effective, these approaches are often prohibitively costly and highly dependent on precise climate and performance data. By contrast, the Global South, confronted with resource constraints, the rapid expansion of informal settlements, and challenging climatic conditions, has relied more extensively on vernacular solutions such as wind catchers, traditional shading systems, and locally sourced materials. Integrating these two trajectories points toward the development of "hybrid" building models that combine the affordability of Global South practices with the efficiency and durability characteristic of the Global North. Empirical studies demonstrate that combining vernacular passive cooling techniques with modern building technologies can reduce indoor temperatures by up to 8–12 °C and lower cooling energy demand by more than 50%, particularly in hot and arid climates [31, 56, 74]. Pilot projects that combine local materials with modern technologies, together with the formulation of integrated design guidelines for engineers and architects, represent concrete strategies for operationalizing this objective. Such strategies demonstrate adaptability across both contexts, ranging from informal settlements in the Global South to high-risk urban neighborhoods in the Global North.

At the design level (Fig. 10), the Global North has accumulated extensive experience in developing urban green infrastructure, permeable surfaces, and vegetative cover, efforts that are typically underpinned by large-scale investments and robust institutional frameworks. By contrast, the Global South, drawing on indigenous vegetation, natural landscapes, and traditional garden practices, has developed low-cost, participatory solutions. Although relatively simple, these solutions yield significant benefits in mitigating urban heat and managing stormwater runoff. Hybrid green infrastructure strategies have been shown to measurably reduce surface temperatures and improve stormwater management, contributing to lower flood risks and enhanced thermal comfort in dense urban environments [48, 49]. An integrative approach at this level entails the development of hybrid green infrastructure that combines the technological and engineering capacities of the Global North with the flexibility and ecological sustainability characteristic of the Global South. Examples of tangible actions include implementing green roofs and walls with indigenous plant species, as well as piloting stormwater management initiatives in high-risk neighborhoods.

At the policy level (Fig. 11), the Global North is distinguished by revised building codes, climate-resilient zoning regulations, and advanced insurance instruments. These instruments provide order and predictability, yet they often lack flexibility in responding to diverse social realities. In the Global South, flexible approaches have emerged through the integration of informal settlements into formal planning and the adoption of land-based policies explicitly informed by climate risk. Integrating these

Table 1 Hybrid strategies for climate-resilient housing: integrating Global North and Global South approaches

Level	Global North strategies	Global South strategies	Proposed integrative strategy	Proposed actions
Physical	• Enhanced building envelopes and advanced thermal insulation [42] • Flood-resilient and amphibious housing models [37] • Integration of green roofs and facades [47, 30] • Deployment of optimized solar energy systems and high-efficiency HVAC technologies [42]	• Natural ventilation and traditional wind-catcher systems; • Use of local materials with high thermal mass [56] • Passive cooling design strategies [84]	Hybrid architectural models that combine indigenous design principles with modern technologies to deliver cost-effective and efficient housing solutions	• Design housing units using local materials (e.g., brick, compressed earth, local stone) reinforced with advanced insulation and solar panels • Develop design guidelines for engineers and architects specifying how modern technologies can be integrated with vernacular features (e.g., wind catchers, traditional shading devices)
Design	• Urban green infrastructure and permeable surfaces [49] • Vegetation cover and urban cooling systems [41]	• Use of indigenous vegetation for heat reduction [84] • Restoration of natural landscapes and traditional gardens	Combine advanced green infrastructure with the restoration of landscapes and indigenous vegetation to simultaneously address heat stress and flooding	• Implement rooftop and vertical greening in urban buildings while promoting native plant species with low irrigation and maintenance needs; • Pilot stormwater management projects in high-risk neighborhoods using permeable surfaces and natural drainage systems adapted to local climates

(continued)

Table 1 (continued)

Level	Global North strategies	Global South strategies	Proposed integrative strategy	Proposed actions
Policy	• Climate-responsive building codes and zoning [41] • Updated legal frameworks and flood-risk maps [55] • Advanced insurance tools and multi-level governance systems	• Integration of informal settlements into formal planning systems [17, 57] • Land-use policies oriented toward climate-risk reduction [59]	Blend institutional rigor with social flexibility to develop policies that are both safe and equitable	• Revise building codes to ensure compliance with safety standards while recognizing and legitimizing indigenous building practices • Establish joint committees within municipalities or housing ministries to exchange insurance and legal experiences and translate them into locally tailored policies
Economic	• Green loans and tax incentives [51] • Climate insurance and innovative financial instruments [78, 28]	• Microcredit and pro-poor subsidies [17] • Community-based financing mechanisms [59]	Design hybrid financial instruments such as "green microloans" that merge the accessibility of microfinance with the sustainability goals of green finance	• Launch green microloan programs with international institutional backing; • Redefine the role of global organizations in supporting climate-resilient housing projects in the Global South

(continued)

Table 1 (continued)

Level	Global North strategies	Global South strategies	Proposed integrative strategy	Proposed actions
Social	• Multi-stakeholder engagement and multi-level governance [78, 28] • Institutional capacity building and participatory planning [87]	• Participatory flood mapping [38] • Community-based upgrading of informal settlements [58] • Local capacity building and social cohesion initiatives [22, 69–71]	Link multi-level governance with community participation to generate hybrid governance models that combine social legitimacy with institutional effectiveness	• Organize joint workshops between formal institutions and local residents for the co-design and implementation of housing upgrading projects • Establish local–urban councils where community representatives and government officials make joint decisions on upgrading strategies
Knowledge and Data	• Data-driven modeling and scenario analysis [87, 30] • Climate monitoring and micro-data collection [88]	• Revitalization of indigenous knowledge and lived experiences [75, 33] • Microclimatic and behavioral data [85]	Integrate advanced modeling and monitoring systems with indigenous and experiential knowledge to generate more precise assessments and reduce the risks of maladaptation	• Establish local–regional monitoring systems to collect fine-grained data (e.g., neighborhood-level temperature, rainfall runoff) and integrate them with advanced climate models • Create knowledge-exchange centers linking universities and research institutes in the Global North and Global South to facilitate the transfer of technology alongside the sharing of vernacular practices and passive design strategies

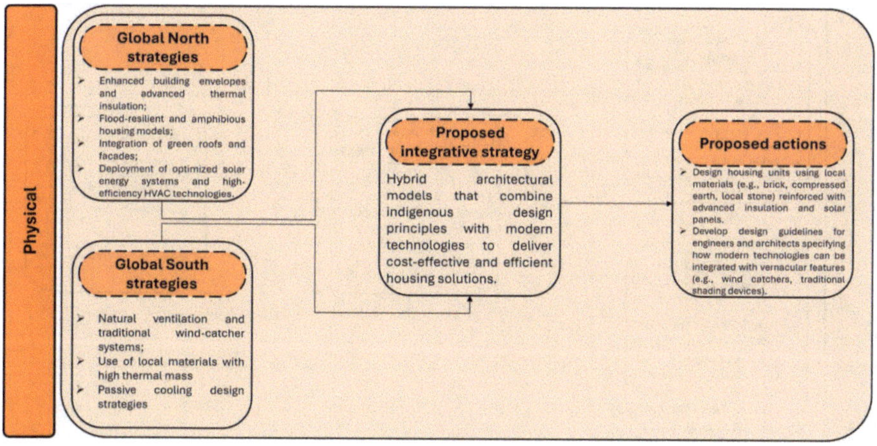

Fig. 9 Synergizing Global North and Global South Strategies to Achieve the Proposed Integrative Strategy and Operational Actions in the Physical Dimension

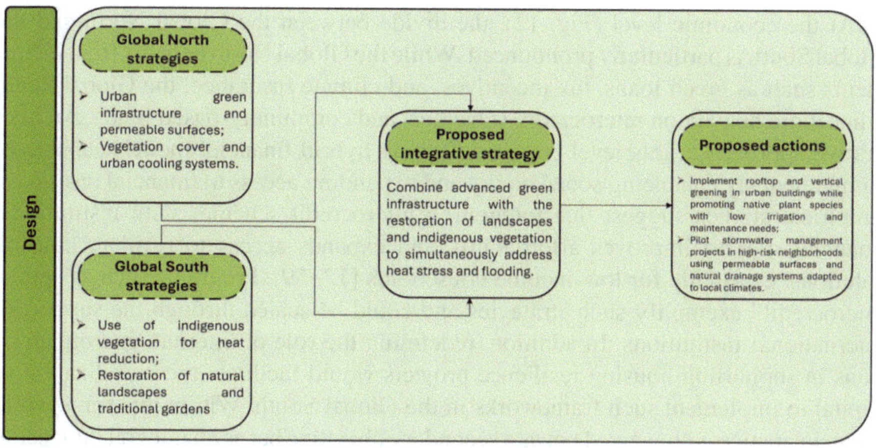

Fig. 10 Synergizing Global North and Global South Strategies to Achieve the Proposed Integrative Strategy and Operational Actions in the Design Dimension

two pathways suggests the creation of dual institutional frameworks that, on the one hand, uphold the safety standards of the Global North while, on the other, recognizing the social realities of the Global South. Evidence from post-disaster housing and climate adaptation programs indicates that flexible regulatory frameworks, when combined with formal safety standards, can significantly improve reconstruction efficiency and regulatory compliance, particularly in low-income and informal contexts [59, 71]. Operational strategies in this domain include revising building regulations to accommodate indigenous materials and practices, and establishing joint local committees to facilitate the exchange of insurance and legal expertise.

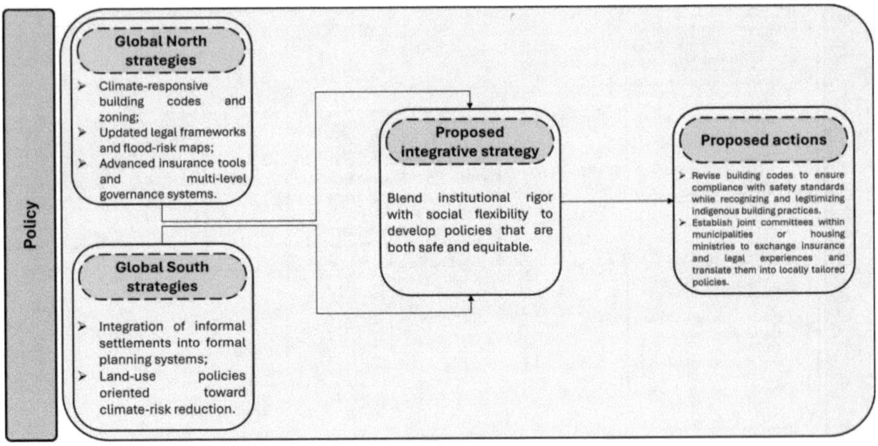

Fig. 11 Synergizing Global North and Global South Strategies to Achieve the Proposed Integrative Strategy and Operational Actions in the Policy Dimension

At the economic level (Fig. 12), the divide between the Global North and the Global South is particularly pronounced. While the Global North benefits from instruments such as green loans, tax incentives, and climate insurance, the Global South relies more heavily on microcredit, subsidies, and community-based funds. An integrative approach at this level entails designing hybrid financial models capable of simultaneously enhancing social equity and expanding access to financial resources. Empirical studies suggest that integrating microcredit schemes with institutional financial support improves affordability and expands access to resilient housing solutions, especially for low-income households [17, 59]. Programs such as "green microcredit" exemplify such strategies and could be scaled through the support of international institutions. In addition, redefining the role of international organizations in supporting housing resilience projects would facilitate the mobilization of capital to implement such frameworks in the Global South. Within this framework, financial justice is conceived not as a secondary objective but as an integral dimension of climate resilience.

At the social level (Fig. 13), the Global North offers valuable experience in multi-level governance, stakeholder participation, and institutional planning, while the Global South has relied more heavily on participatory mapping, settlement upgrading, and strengthening social cohesion. Integrating these two pathways can foster hybrid governance models that combine social legitimacy with institutional effectiveness. Quantitative and comparative assessments from post-disaster housing programs show that participatory and multi-level governance models achieve higher uptake, stronger community ownership, and improved long-term maintenance outcomes compared to purely top-down approaches [69, 70]. Practical strategies include joint workshops between formal institutions and local communities, as well as the establishment of local–urban councils for collective decision-making, both

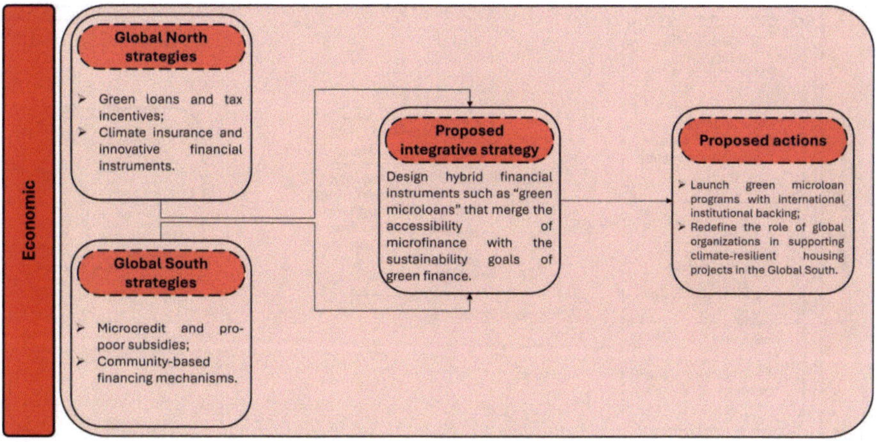

Fig. 12 Synergizing Global North and Global South Strategies to Achieve the Proposed Integrative Strategy and Operational Actions in the Economic Dimension

of which can strengthen trust between state actors and communities. Such governance not only enhances the quality of decision-making but also facilitates the more sustainable implementation of climate adaptation measures.

Finally, at the level of knowledge and data (Fig. 14), the Global North is distinguished by advanced data-driven modeling tools, scenario analysis, and climate monitoring, while the Global South relies more heavily on indigenous knowledge, lived experiences, and microclimatic observations. Integrating these two pathways can generate more robust assessments and foster more equitable decision-making. Studies indicate that integrating local microclimatic observations with advanced

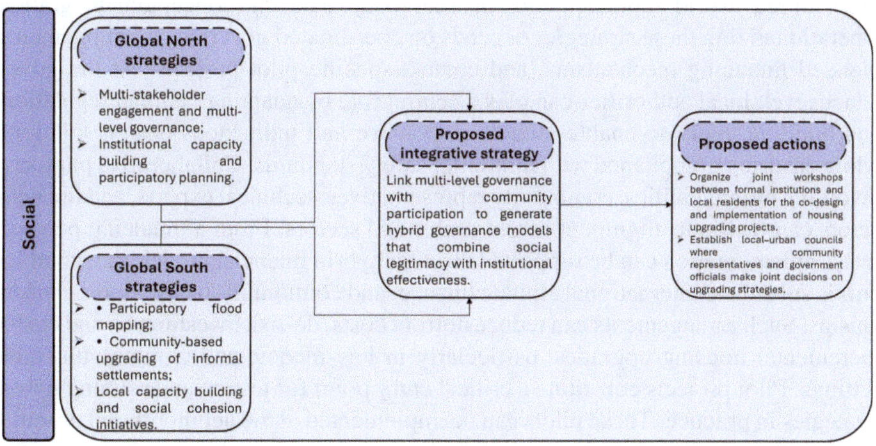

Fig. 13 Synergizing Global North and Global South Strategies to Achieve the Proposed Integrative Strategy and Operational Actions in the Social Dimension

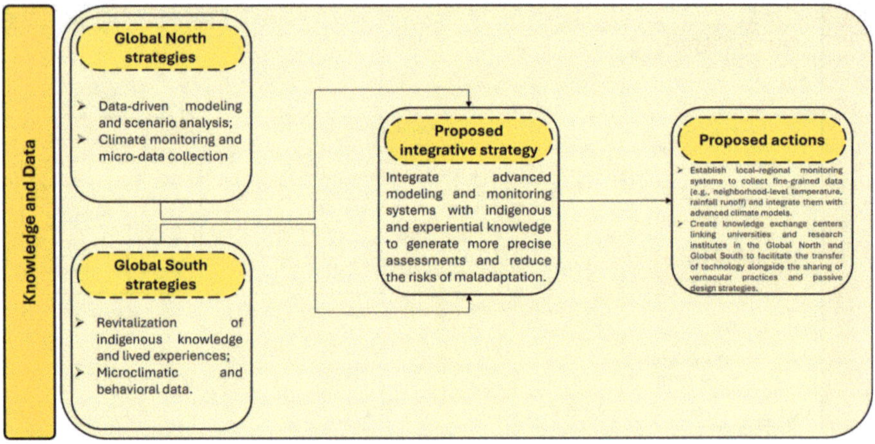

Fig. 14 Synergizing Global North and Global South Strategies to Achieve the Proposed Integrative Strategy and Operational Actions in the Knowledge and Data Dimension

climate modeling enhances the reliability of risk assessments and supports more context-sensitive housing design decisions [85, 88]. Operational strategies at this level include developing local–regional monitoring systems that gather micro-scale data, such as temperature and runoff, and integrate them with advanced climate models. In addition, establishing knowledge-exchange centers between universities and research institutes in the Global North and South would enable the reciprocal transfer of technology and experiential knowledge. Such initiatives elevate knowledge from a purely scientific domain to an applied one, grounding decision-making simultaneously in both local realities and global perspectives.

Taken together, the multi-level hybrid strategies outlined above require clear pathways for real-world implementation in order to move beyond conceptual integration. Operationalizing these strategies depends on coordinated governance arrangements, blended financing mechanisms, and context-specific pilot projects. At the governance level, local authorities can play a central role by adapting national regulations and building codes to enable climate-responsive and indigenous design solutions, while ensuring compliance with minimum safety standards. Collaborative platforms involving municipalities, community representatives, technical experts, and financial actors can facilitate alignment across scales and sectors. From a financing perspective, implementation can be supported through hybrid financial models that combine public subsidies, international climate finance, and community-based funding mechanisms. Such arrangements can reduce upfront costs, de-risk investments, and enable incremental housing upgrades, particularly in low-income and climate-vulnerable settings. Pilot projects constitute a critical entry point for testing and refining hybrid strategies in practice. These pilots can be implemented at the neighborhood or settlement scale, focusing on areas exposed to flooding or extreme heat, and can integrate passive design measures, renewable energy systems, and hybrid green infrastructure. Systematic monitoring of pilot outcomes, such as thermal performance, energy use, affordability, and community participation, can generate transferable lessons and support the scaling-up of successful approaches (Fig. 15).

The proposed hybrid framework leads to several overarching takeaways:

- Hybrid housing resilience strategies are most effective when technological and institutional capacities are combined with indigenous knowledge and community-based practices.
- Operationalizing hybrid strategies requires coordinated governance mechanisms, blended financing models, and context-specific pilot projects.
- Hybrid approaches enhance adaptability, affordability, and social legitimacy across diverse climatic and socio-economic contexts.
- Moving beyond binary Global North–Global South models enables mutual learning and supports more equitable and scalable climate-resilient housing solutions.

5 Conclusion

5.1 Summary

This study demonstrates that climate change exerts multidimensional and differentiated impacts on the housing sector in both the Global North and the Global South. Whereas in the Global North challenges are largely associated with rising costs, financial risks, and demands for resilient infrastructure, in the Global South they encompass more acute dimensions, including widespread housing destruction, large-scale displacement, and increased mortality. The systematic review of the literature revealed a diverse range of strategies to strengthen housing resilience, ranging from advanced technologies and green infrastructure in the Global North to vernacular architecture, community participation, and microfinance mechanisms in the Global South. However, each of these approaches carries inherent limitations and cannot, in isolation, address the complexities of the climate crisis.

The comparative framework advanced in this study is grounded in the creative integration of experiences from the Global North and Global South, generating "hybrid" and context-sensitive models for enhancing housing resilience. Overall, the central message of this study is that housing resilience in the age of climate change can only be realized in a sustainable and equitable manner if the experiences of the two worlds are mobilized not in opposition but in complementarity. By integrating technological, institutional, and social capacities, this comparative framework provides a practical model for policymakers, planners, and designers, enabling them not only to reduce climate risks but also to advance social justice and enhance quality of life in some of the world's most vulnerable regions.

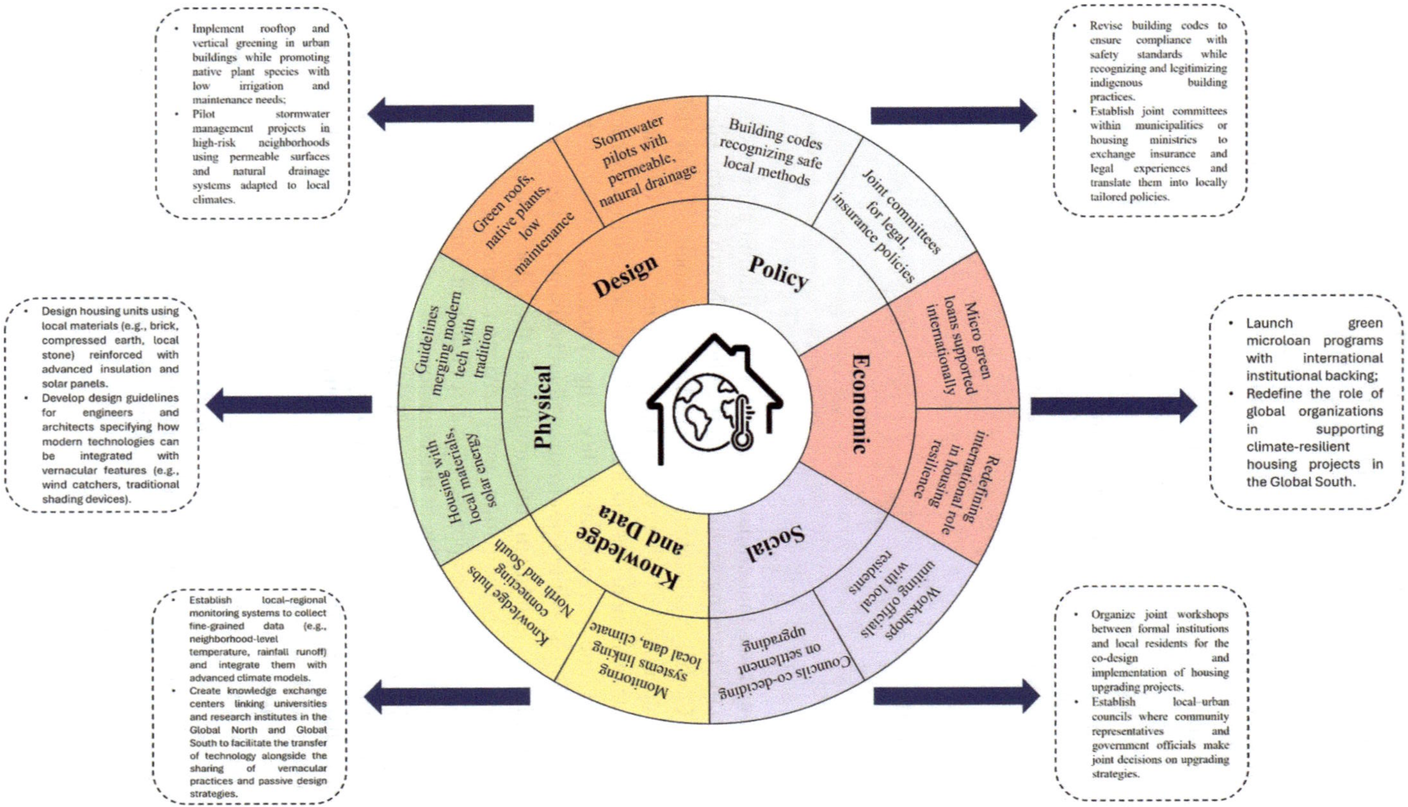

Fig. 15 Pathways toward climate-resilient housing

5.2 Theoretical and Practical Implications

From a theoretical perspective, this study integrates technological and vernacular approaches to advance a comparative framework for analyzing housing resilience. This framework can serve as a foundation for future research on housing policy and planning, built environment resilience, climate justice, urban governance, and sustainable development. The framework conceives housing not merely as a physical structure but as a social, economic, and cultural institution that is continuously shaped through its interaction with climatic forces.

From a practical perspective, the findings yield several clear implications for policymakers and urban planners:

- The integration of indigenous knowledge with modern technologies can yield cost-effective and efficient housing models.
- Dual institutional policies must simultaneously uphold the safety standards of the Global North and acknowledge the social flexibility of the Global South.
- Hybrid financial instruments, such as green microcredit, can advance social justice while expanding access to financial resources.
- Hybrid governance, through the joint participation of formal institutions and local communities, can enhance social legitimacy and institutional effectiveness.

5.3 Limitations and Recommendations for Future Research

Despite its broad scope, this study is subject to several limitations. First, the study's reliance on English-language sources may have led to the omission of valuable local experiences documented in other languages. Second, the exclusive reliance on the Web of Science (WoS) database may have excluded relevant grey literature and studies not indexed within this platform. Third, the dataset was largely composed of case studies, which limits the generalizability of the findings. Finally, given the wide institutional and cultural variations across contexts, the practical implementation of the proposed framework will require careful contextualization and field-based validation.

Building on these limitations, several directions for future research can be identified:

- Conduct further comparative studies across diverse regions to illuminate both common patterns and contextual differences.
- Undertake field-based research with active participation of local communities to evaluate the real-world effectiveness of hybrid strategies.
- Develop quantitative tools and advanced modeling in combination with indigenous knowledge and microclimatic data to enable more precise assessments of proposed strategies.
- Formulate multi-scalar strategies to enhance the resilience and adaptability of housing units, buildings, and residential neighborhoods.

- Strengthen investigation into the role of international organizations and global financing mechanisms in supporting housing resilience in the Global South.

Declaration of Generative AI During the preparation of this manuscript, the author(s) used the Gemini tool to assist with initial language translation into English. Following this, the author(s) thoroughly reviewed, revised, and edited the content to ensure accuracy, clarity, and academic integrity. The author(s) take full responsibility for the content of the final submitted version.

References

1. Ahmadi Dehrashid P, Pourahmad A, Mansourian H, Khavarian-Garmsir AR (2025) Genealogy of the concept of healthy housing. In: Almulhim AI, Leal Filho W, Sharifi A (eds) Urbanization, climate change, and health. Climate Change Management. Springer, Cham. https://doi.org/10.1007/978-3-031-99132-5_9
2. Bezgrebelna M, McKenzie K, Wells S, Ravindran A, Kral M, Christensen J, Stergiopoulos V, Gaetz S, Kidd SA (2021) Climate change, weather, housing precarity, and homelessness: a systematic review of reviews. Int J Environ Res Public Health 18(11):5812. https://doi.org/10.3390/ijerph18115812
3. Hales S, Baker M, Howden-Chapman P, Menne B, Woodruff R, Woodward A (2007) Implications of global climate change for housing, human settlements and public health. Rev Environ Health 22(4):295–302. https://doi.org/10.1515/REVEH.2007.22.4.295
4. Vera F, Uribe MC, Del Castillo S (2023) Climate action and the Paris agreement: the role of cities in Latin America and the Caribbean. Inter-American Development Bank. https://doi.org/10.18235/0004837
5. D'Alessandro D, Appolloni L (2020) Housing and health: an overview. Ann Hyg Prev Community Med 32(5 Supple 1):17–26. https://iris.uniroma1.it/handle/11573/1455941
6. Krieger J, Higgins DL (2002) Housing and health: time again for public health action. Am J Public Health 92(5):758–768. https://doi.org/10.2105/AJPH.92.5.758
7. WHO (2018) Housing and Health Guidelines. https://www.who.int/publications/i/item/9789241550376
8. Aliu IR (2022) Sustainable housing development dynamics in the Global South: refl ections on theories, strategies and constraints. Bull Geogr. Socio-Econ Ser 56:83–100. https://doi.org/10.12775/bgss-2022-0014
9. Chu E, Anguelovski I, Roberts D (2017) Urban climate adaptation in the global South. In: The Routledge companion to planning in the Global South. Routledge. https://doi.org/10.4324/9781317392842-14
10. Van Tonder F, Rwelamila PD (2024) A comparison of low-cost housing units for varying climatic regions in South Africa: a knowledge management approach. Smart Sustain Built Environ 13(5):1310–1329. https://doi.org/10.1108/SASBE-11-2022-0237
11. Czischke D, Ayala A (2021) Housing in the global north and the global south. Companion to urban and regional studies, pp. 579–604. https://doi.org/10.1002/9781119316916.ch27
12. Almulhim AI, Alverio GN, Sharifi A, Shaw R, Huq S, Mahmud MJ, Ahmad S, Abubakar IR (2024) Climate-induced migration in the Global South: an in depth analysis. Npj Clim Action 3(1):47. https://doi.org/10.1038/s44168-024-00133-1
13. Khavarian-Garmsir AR, Pourahmad A, Hataminejad H, Farhoodi R (2019) Climate change and environmental degradation and the drivers of migration in the context of shrinking cities: a case study of Khuzestan province, Iran. Sustain Cities Soc 47:101480. https://doi.org/10.1016/j.scs.2019.101480

14. Ngcamu BS (2023) Climate change effects on vulnerable populations in the Global South: a systematic review. Nat Hazards 118(2):977–991. https://doi.org/10.1007/s11069-023-06070-2
15. Ruíz MA, Mack-Vergara YL (2023) Resilient and sustainable housing models against climate change: a review. Sustainability 15(18):13544. https://doi.org/10.3390/su151813544
16. Hernández D (2022) Climate justice starts at home: building resilient housing to reduce disparate impacts from climate change in residential settings. Am J Public Health 112(1):66–68. https://doi.org/10.2105/AJPH.2021.306611
17. Adetokunbo I, Emeka M (2015) Urbanization, housing, homelessness and climate change. J Des Built Environ 15(2):15–28. https://doi.org/10.22452/jdbe.vol15no2.3
18. Rubio-Bellido C, Pulido-Arcas JA, Cabeza-Lainez JM (2015) Adaptation strategies and resilience to climate change of historic dwellings. Sustainability 7(4):3695–3713. https://doi.org/10.3390/su7043695
19. Mavrogianni A, Taylor J, Davies M, Thoua C, Kolm-Murray J (2015) Urban social housing resilience to excess summer heat. Build Res Inf 43(3):316–333. https://doi.org/10.1080/09613218.2015.991515
20. Porter L, Steele W, Stone W (2018) Housing and resilience—when, for whom and for what? A critical agenda. Hous Theory Soc. https://doi.org/10.1080/14036096.2018.1492964
21. Hage S, Vegas López-Manzanares F, Mileto C, Hollermann S (2025) Designing for resilience: housing needs and climate perceptions in rural Siaya County, Kenya. Buildings 15(16):2947. https://doi.org/10.3390/buildings15162947
22. Chohan AH, Awad J, Jung C, Sher BK (2024) Enhancing climate resilience against flooding in housing design through synergistic strategies in Pakistan. Future Cities Environ 10(1). https://doi.org/10.5334/fce.226
23. Moher D, Liberati A, Tetzlaff J, Altman DG (2009) Preferred reporting items for systematic reviews and meta-analyses: the PRISMA statement. BMJ 339(1):b2535–b2535. https://doi.org/10.1136/bmj.b2535
24. Page MJ, McKenzie JE, Bossuyt PM, Boutron I, Hoffmann TC, Mulrow CD, Shamseer L, Tetzlaff JM, Akl EA, Brennan SE, Chou R, Glanville J, Grimshaw JM, Hróbjartsson A, Lalu MM, Li T, Loder EW, Mayo-Wilson E, McDonald S, Moher D (2021) The PRISMA 2020 statement: an updated guideline for reporting systematic reviews. Syst Rev 10(1):89. https://doi.org/10.1186/s13643-021-01626-4
25. Assarroudi A, Heshmati Nabavi F, Armat MR, Ebadi A, Vaismoradi M (2018) Directed qualitative content analysis: the description and elaboration of its underpinning methods and data analysis process. J Res Nurs 23(1):42–55. https://doi.org/10.1177/1744987117741667
26. Esmaeili M, Sheydayi A, Jamalabadi F (2025) A systematic review of qualitative and quantitative content analysis applications in urban planning research: proposing a mixed method approach. J Urban Plann Dev 151(3). https://doi.org/10.1061/JUPDDM.UPENG-5080
27. Hertin J, Berkhout F, Gann D, Barlow J (2003) Climate change and the UK house building sector: perceptions, impacts and adaptive capacity. Build Res Inf 31(3–4):278–290. https://doi.org/10.1080/0961321032000097683
28. Storey B, Noy I, Owen S, Townsend W, Kerr S, Salmon R, Middleton D, Filippova O, James V (2017) Insurance, Housing and Climate Adaptation: current knowledge and future research. https://hdl.handle.net/2292/39571
29. Kahn ME (2016) The climate change adaptation literature. Rev Environ Econ Policy. https://doi.org/10.1093/reep/rev023
30. Roders M, Straub A, Visscher H (2013) Evaluation of climate change adaptation measures by Dutch housing associations. Struct Surv 31(4):267–282. https://doi.org/10.1108/SS-01-2013-0009
31. Hashemi A (2016) Climate resilient low-income tropical housing. Energies 9(6):468. https://doi.org/10.3390/en9060468
32. Chenary K, Abdi MH (2024) Cities, arid climates and shading: persian vernacular building responses. In: Cheshmehzangi A, Roaf S (eds) Persian vernacular architecture. Urban sustainability. Springer, Singapore. https://doi.org/10.1007/978-981-96-1116-4_11

33. Roaf S, Cheshmehzangi A (2024) Persian vernacular buildings—lessons on resilient design. In: Cheshmehzangi A, Roaf S (eds) Persian vernacular architecture. Urban sustainability. Springer, Singapore. https://doi.org/10.1007/978-981-96-1116-4_1
34. Gorji M-A, Hadavi N, Nahavandi N (2023) Improving urban resilience against climate change through government tax policies to housing companies: a game-theoretic approach. Urban Clim 49:101565. https://doi.org/10.1016/j.uclim.2023.101565
35. New America (2022) Housing and climate change in the United States: introduction: climate impacts on U.S. housing security. https://www.newamerica.org/future-land-housing/reports/housing-and-climate-change-in-the-united-states/introduction-climate-impacts-on-us-housing-security/
36. Government of Malawi (2023) Malawi 2023 tropical cyclone Freddy post-disaster needs assessment. Government of Malawi, Lilongwe. https://reliefweb.int/attachments/6e65f674-a5f8-4b18-a913-a9aa141a7c04/Tropical%20Cyclone%20Freddy%20Post%20Disaster%20Needs%20Assessment%20Government%20of%20Malawi%20%20April%202023-12_05_2023.pdf
37. Rydborg M, Lauring M, Brunsgaard C (2019) Vulnerabilities and resilience in Danish housing stock: a comparative study of architectural answers to climate change in Danish housing in relation to other oceanic climates. E3S Web Conf 111:03068. https://doi.org/10.1051/e3sconf/201911103068
38. Tschakert P, Dietrich KA (2010) Anticipatory learning for climate change adaptation and resilience. Ecol Soc 15(2):art11. https://doi.org/10.5751/ES-03335-150211
39. Daoudi NS, Mestoul D, Lamraoui S, Boussoualim A, Adolphe L, Bensalem R (2019) Vernacular architecture in arid climates: adaptation to climate change. In: Guedes M, Cantuaria G (eds) Bioclimatic architecture in warm climates. Springer, Cham. https://doi.org/10.1007/978-3-030-12036-8_4
40. Henna K, Saifudeen A, Mani M (2021) Resilience of vernacular and modernising dwellings in three climatic zones to climate change. Sci Rep 11(1):9172. https://doi.org/10.1038/s41598-021-87772-0
41. De Bruin K, Dellink RB, Ruijs A, Bolwidt L, van Buuren A, Graveland J, De Groot RS, Kuikman PJ, Reinhard S, Roetter RP (2009) Adapting to climate change in The Netherlands: an inventory of climate adaptation options and ranking of alternatives. Clim Change 95(1):23–45. https://doi.org/10.1007/s10584-009-9576-4
42. Ren ZE, Chen ZD, Wang XM (2011) Climate change adaptation pathways for Australian residential buildings. Build Environ 46(11):2398–2412. https://doi.org/10.1016/j.buildenv.2011.05.022
43. Bondinuba F, Jones CA, Akinbogun S (2025) Building equitable and resilient communities in the Global South through housing justice and climate justice. https://doi.org/10.2139/ssrn.5260550
44. Das DK (2025) Resilient and sustainable urban infrastructure in the Global South: strategies to address climate change-linked vulnerabilities. Urban Plann 10. https://doi.org/10.17645/up.10037
45. Díaz CG, Zambrana-Vasquez D, Bartolomé C (2024) Building resilient cities: a comprehensive review of climate change adaptation indicators for urban design. Energies 17(8):1959. https://doi.org/10.3390/en17081959
46. Nordkirche (2022) Pakistan: North Church provides 15,000 euros and asks for donations. https://www.nordkirche.de/nachrichten/nachrichten-detail/nachricht/pakistan-nordkirche-stellt-15000-euro-bereit-und-bittet-um-spenden
47. Boezeman D, de Vries T (2019) Climate proofing social housing in the Netherlands: toward mainstreaming? J Environ Plann Manag 62(8):1446–1464. https://doi.org/10.1080/09640568.2018.1510768
48. Rawlins BG, Harris J, Price S, Bartlett M (2015) A review of climate change impacts on urban soil functions with examples and policy insights from England, UK. Soil Use Manag 31(S1):46–61. https://doi.org/10.1111/sum.12079

49. Taleghani M, Kleerekoper L, Tenpierik M, van den Dobbelsteen A (2015) Outdoor thermal comfort within five different urban forms in the Netherlands. Build Environ 83:65–78. https://doi.org/10.1016/j.buildenv.2014.03.014
50. Green SD (2014) Building resilient communities in the wake of climate change while keeping affordable housing safe from sea changes in nature and policy. Washburn LJ 54:527
51. Roders M, Straub A (2015) Assessment of the likelihood of implementation strategies for climate change adaptation measures in Dutch social housing. Build Environ 83:168–176. https://doi.org/10.1016/j.buildenv.2014.07.014
52. Adeyemi AB, Ohakawa TC, Okwandu AC, Iwuanyanwu O, Ifechukwu G-O (2024) Affordable housing and resilient design: preparing low-income housing for climate change impacts. J Sustain Architect 12(2):45–60
53. Li A, Toll M, Chapman R, Howden-Chapman P, Hernández D, Samuelson H, Woodward A, Bentley R (2025) Housing at the intersection of health and climate change. Lancet Public Health. https://doi.org/10.1016/S2468-2667(25)00141-0
54. Hertman L (2024) Adaptation of the housing and utilities sector of belarus to climate change. Reliability 19:654. https://openurl.ebsco.com/contentitem/gcd:182083535?sid=ebsco:plink:crawler&id=ebsco:gcd:182083535
55. McKay R, Martin G, Goemans M (2023) Politics, housing and climate adaptation in Ottawa, Canada. Can J Urban Res 32(2):59. https://openurl.ebsco.com/contentitem/gcd:178539684?sid=ebsco:plink:crawler&id=ebsco:gcd:178539684
56. Triana MA, Lamberts R, Sassi P (2018) Should we consider climate change for Brazilian social housing? Assessment of energy efficiency adaptation measures. Energy Build 158:1379–1392. https://doi.org/10.1016/j.enbuild.2017.11.003
57. Régis Araújo R, Palazzi Perez L, Trigueiro Morais M, Lira Anjos K (2024) Emergência climática e precariedade habitacional urbana: diretrizes para adaptação climática no bairro São José, João Pessoa—PB. Desenvolvimento e Meio Ambiente. https://doi.org/10.5380/dma.v63i0.87786
58. Diep L (2023) Climate adaptation through housing: examples of effective interventions for informal settlements. https://www.habitat.org/sites/default/files/documents/climate-adaptation-through-housing-report.pdf
59. Brown A, Dayal A, del Rio CR (2012) From practice to theory: emerging lessons from Asia for building urban climate change resilience. Environ Urban 24(2):531–556. https://doi.org/10.1177/0956247812456490
60. Ruíz-González MA, Mack-Vergara YL (2022) Indicadores de resiliencia y sostenibilidad para la vivienda urbana panameña frente al cambio climático. Revista Hábitat Sustentable 08–25. https://doi.org/10.22320/07190700.2022.12.02.01
61. Tea After Twelve (n.d.) Brazil without favelas. Issue 01—Overview, Chapter 4. http://www.tea-after-twelve.com/all-issues/issue-01/issue-01-overview/chapter4/brazil-without-favelas/index.html
62. Roders M, Straub A, Visscher H (2012) Awareness of climate change adaptations among Dutch housing associations. Open House Int 37(4):61–71. https://doi.org/10.1108/OHI-04-2012-B0007
63. Al-Hassawi OD, Drake D (2023) Innovations in passive downdraft cooling performance evaluation methods: design and construction of a novel environmental test chamber. Energies 16(11):4371. https://doi.org/10.3390/en16114371
64. Kokatnur T, Ferreira S, Akkoç BK, Markarian E, Nojedehi P, Qiblawi S, Sewraj K, Gunay B, O'Brien W, Papineau M, Schweiker M, Harputlugil GU, Azar E (2025) A review of passive design strategies and their effect on thermal resilience in low-income households. Energy Build 348:116508. https://doi.org/10.1016/j.enbuild.2025.116508
65. Monge Palma R, Castro Medina D, Guerrero Delgado Mc, Sánchez Ramos J, Montero-Gutiérrez P, Álvarez Domínguez S (2024) Enhancing the building resilience in a changing climate through a passive cooling roof: a case study in Camas (Seville, Spain). Energy Build 321:114680. https://doi.org/10.1016/j.enbuild.2024.114680

66. Erlwein S, Meister J, Wamsler C, Pauleit S (2023) Governance of densification and climate change adaptation: How can conflicting demands for housing and greening in cities be reconciled? Land Use Policy 128:106593. https://doi.org/10.1016/j.landusepol.2023.106593
67. Hernández EH (2018) Prototipo de vivienda vertical social sustentable, enfoque en resistencia al cambio climático. Revista INVI 33(92):213–237. https://doi.org/10.4067/S0718-83582018000100213
68. Song X, Ye C (2017) Climate change adaptation pathways for residential buildings in southern China. Energy Procedia 105:3062–3067. https://doi.org/10.1016/j.egypro.2017.03.635
69. Haigh R, Hettige S, Sakalasuriya M, Vickneswaran G, Weerasena LN (2016) A study of housing reconstruction and social cohesion among conflict and tsunami affected communities in Sri Lanka. Disaster Prev Manag 25(5):566–580. https://doi.org/10.1108/DPM-04-2016-0070
70. Hayles CS (2010) An examination of decision making in post disaster housing reconstruction. Int J Disaster Resilience Built Environ 1(1):103–122. https://doi.org/10.1108/17595901011026508
71. Tran TA (2015) Post-disaster housing reconstruction as a significant opportunity to building disaster resilience: a case in Vietnam. Nat Hazards 79(1):61–79. https://doi.org/10.1007/s11069-015-1826-3
72. Afsahhosseini F (2024) The impact of Iran's urban heritage on sustainability, climate change and carbon zero. Environ Dev Sustain 27(6):12351–12391. https://doi.org/10.1007/s10668-023-04434-z
73. Hassani EK (2024) Preserving the past, designing the future: unveiling climate-resilient design in persian vernacular architecture for modern sustainability. In: Cheshmehzangi A, Roaf S (eds) Persian vernacular architecture. Urban sustainability. Springer, Singapore. https://doi.org/10.1007/978-981-96-1116-4_15
74. Khalili M, Amindeldar S (2014) Traditional solutions in low energy buildings of hot-arid regions of Iran. Sustain Cities Soc 13:171–181. https://doi.org/10.1016/j.scs.2014.05.008
75. Cheshmehzangi A, Roaf S (2024) Persian vernacular architecture for climate resilient design thinking: looking back, looking forward. In: Cheshmehzangi A, Roaf S (eds) Persian vernacular architecture. Urban sustainability. Springer, Singapore. https://doi.org/10.1007/978-981-96-1116-4_17
76. Green Prophet (2010) The Wind Catchers of Yazd, Iran. https://www.greenprophet.com/2010/05/wind-catchers-iran-yazd/
77. Rözer V, Surminski S (2021) Current and future flood risk of new build homes across different socio-economic neighbourhoods in England and Wales. Environ Res Lett 16(5):054021. https://doi.org/10.1088/1748-9326/abec04
78. Rözer V, Surminski S (2021) New build homes, flood resilience and environmental justice-current and future trends under climate change across england and wales. Grantham Res Inst Clim Change Environ. https://doi.org/10.5194/egusphere-egu21-12494
79. Shearer H, Coiacetto E, Dodson J, Taygfeld P (2016) How the structure of the Australian housing development industry influences climate change adaptation. Hous Stud 31(7):809–828. https://doi.org/10.1080/02673037.2016.1150430
80. Hayles CS, Dean M (2015) Social housing tenants, climate change and sustainable living: a study of awareness, behaviours and willingness to adapt. Sustain Cities Soc 17:35–45. https://doi.org/10.1016/j.scs.2015.03.007
81. Hegger DLT, Mees HLP, Driessen PPJ, Runhaar HAC (2017) The roles of residents in climate adaptation: a systematic review in the case of the Netherlands. Environ Policy Gov 27(4):336–350. https://doi.org/10.1002/eet.1766
82. Aleksić J, Kosanović S, Tomanović D, Grbić M, Murgul V (2016) Housing and climate change-related disasters: a study on architectural typology and practice. Procedia Eng 165:869–875. https://doi.org/10.1016/j.proeng.2016.11.786
83. Stagrum AE, Andenæs E, Kvande T, Lohne J (2020) Climate change adaptation measures for buildings—a scoping review. Sustainability 12(5):1721. https://doi.org/10.3390/su12051721
84. Baghaiepoor M, Jovanovic G, Stanimirovic M (2019) Climate adapted houses in Iran: hot, cold and humid climate. Facta Universitatis—Ser 17(4):429–443. https://doi.org/10.2298/FUACE180721025B

85. Saifudeen A, Mani M (2024) Adaptation of buildings to climate change: an overview. Front Built Environ 10:1327747. https://doi.org/10.3389/fbuil.2024.1327747
86. Rajkovich NB, Okour Y (2019) Climate change resilience strategies for the building sector: examining existing domains of resilience utilized by design professionals. Sustainability 11(10):2888. https://doi.org/10.3390/su11102888
87. Carriere ZD, Lesnikowski A (2024) The climate change adaptation readiness of co-operative housing in Nova Scotia, Canada. Environ Policy Govern 34(6):724–739. https://doi.org/10.1002/eet.2116
88. Wright A, Venskunas E (2022) Effects of future climate change and adaptation measures on summer comfort of modern homes across the regions of the UK. Energies 15(2):512. https://doi.org/10.3390/en15020512

Technological, Infrastructural, and Systemic Pathways to Climate Resilience

AI-Enabled Emerging Technologies for High-Performance Building Benchmarking: Balance Energy, Comfort, and Health Resilience

Tian Li, Yunyang Ye, and Yi Lu

1 Introduction

The building stock is a major lever for climate action, accounting for approximately 40% of global energy use and associated CO_2 emissions [29]. Reducing this footprint requires addressing both operational and embodied carbon, especially as climate change intensifies. Heatwaves, wildfires, and other extreme events are pushing buildings beyond their limits [40]. Key mitigation strategies span envelope improvements, high-efficiency and electrified building systems, passive design, on-site renewables, and improved commissioning and operation [12, 13]. Among these strategies, energy benchmarking approaches can deliver persistent savings across portfolios while supporting grid flexibility through demand response. When embedded within disclosure policies, codes, and performance standards, energy benchmarking can accelerate progress toward net-zero targets across new construction and, critically, the large existing-building stock [25, 55].

Energy, however, is only one dimension of building performance. As people spend most of their lives indoors, indoor environmental quality (IEQ), including thermal comfort, air quality, lighting, and acoustics, directly affects health, cognition, and productivity [10]. Therefore, high-performance design integrates energy efficiency with robust IEQ strategies, such as adequate ventilation and filtration (particularly

T. Li (✉)
College of Architecture, University of Nebraska—Lincoln, Lincoln, NE, USA
e-mail: tli18@unl.edu

Y. Ye
Department of Civil and Environmental Engineering, Colorado School of Mines, Golden, CO, USA
e-mail: yunyang.ye@mines.edu

Y. Lu
Tepper School of Business, Carnegie Mellon University, Pittsburgh, PA, USA

A. Cheshmehzangi et al. (eds.), *Design Innovation for Climate-Resilient Cities and Built Environments*, Urban Sustainability, https://doi.org/10.1007/978-981-92-2297-1_8

during wildfire smoke or pollution events), moisture and contaminant control, glare-aware daylighting, and thermal monitoring, supported by post-occupancy evaluation [15] (Zhang et al. 2025). The resulting goal is a low-carbon, resilient indoor environment that actively supports human well-being.

At the urban scale, climate resilience depends on how buildings collectively perform during extreme events such as heatwaves, power outages, and air pollution episodes [40]. Buildings that maintain safe temperatures, adequate air quality, and basic functionality without continuous energy input strengthen neighborhood resilience, while poorly performing buildings amplify vulnerability (Zhang et al. 2025). In this context, AI-enabled benchmarking becomes a critical planning tool, enabling cities to identify resilience gaps, prioritize retrofits, and support equitable adaptation across diverse building stocks.

A variety of methods have been developed to benchmark building performance, generally categorized into three groups: (i) model-based simulations, (ii) rating or certification systems, and (iii) data-driven methods [34, 50]. Simulation-based approaches rely on physics-based tools, such as EnergyPlus, DOE-2, and similar software, to estimate baseline energy consumption under assumed operating conditions [46]. While these simulations are useful for exploring end-use loads and testing design alternatives, their accuracy is limited without extensive calibration using actual utility and operational data, which can be time-consuming and resource-intensive.

Point-based rating systems, such as the Leadership in Energy and Environmental Design (LEED) in the United States and the Building Research Establishment Environmental Assessment Methodology (BREEAM) in the United Kingdom, offer a structured framework for evaluating sustainability and promoting high-performance practices [8]. They are widely adopted for certifying sustainable buildings and encouraging compliance with established standards. These systems emphasize prescriptive criteria and aggregate scores, which obscures fine-grained performance variation across buildings and time, while their labor-intensive assessment processes limit scalability [28].

In contrast to simulation and point-based methods, data-driven approaches have gained significant attention in recent years, largely due to rapid advances in ML and AI [57]. These methods leverage large-scale operational data, including utility records, sensor networks, weather information, and even remote sensing imagery, to train predictive models that accurately estimate building performance under diverse operating conditions [34]. Unlike simulation-based or point-based tools, which depend heavily on assumed inputs or rating systems, data-driven approaches learn directly from empirical evidence, making them more adaptable and scalable [4].

Advanced AI algorithms, including ensemble learning, deep neural networks, and hybrid models, can capture complex and nonlinear relationships among building characteristics, occupant behaviors, and environmental variables [45]. This not only improves predictive accuracy but also supports granular insights across multiple time scales, from hourly energy-use patterns to annual carbon footprints (Guo et al. 2025). Furthermore, AI models can be continuously updated with new data streams, ensuring that benchmarks remain current and responsive to changing conditions. When paired

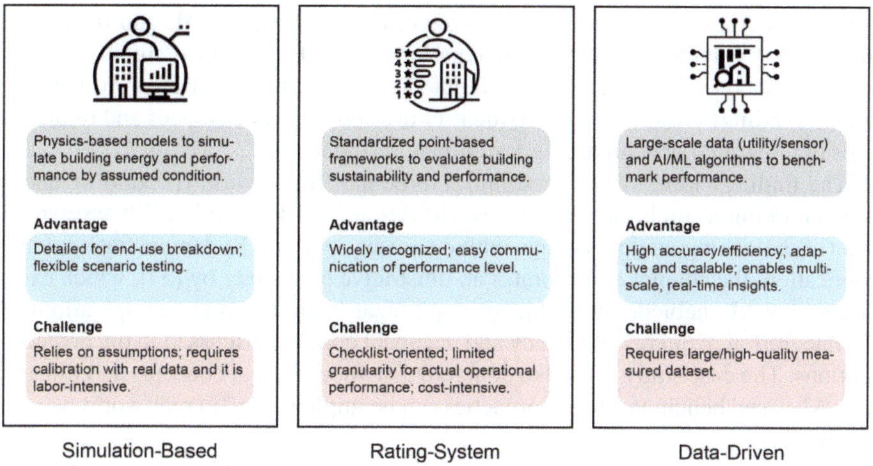

Fig. 1 Definitions and characteristics for simulation-based, rating-system, and data-driven building performance benchmarking approaches

with explainable AI (XAI) techniques, such models provide interpretable outputs that help designers, operators, and policymakers bridge the gap between predictive accuracy and actionable guidance [53]. The definitions and features of the three discussed approaches for performance benchmarking are shown in Fig. 1.

To date, most AI-driven building performance benchmarking studies have focused primarily on either energy loads or IEQ and health outcomes, with limited attention to their integration. This imbalance highlights the need for more holistic approaches that focus on both essential aspects. Accordingly, this chapter introduces advanced AI-powered benchmarking frameworks for high-performance buildings that address both energy performance and IEQ/health dimensions, thereby establishing a more comprehensive foundation for sustainable, occupant-centered design.

2 Methodology

This chapter adopts a narrative-based literature review approach, complemented by an illustrative case study, to explore how AI and related emerging technologies can support high-performance building benchmarking. The chapter is designed to identify advances in energy benchmarking, IEQ, and occupant health, while also highlighting equity considerations.

To investigate the state of research, we conduct a structured narrative review of peer-reviewed studies published primarily since 2015 across Google Scholar, Scopus, and Web of Science. We prioritize studies that apply AI/ML or related data-driven techniques to building performance benchmarking and that report outcomes beyond energy-only metrics, including IEQ, comfort, or health-related indicators. Search strategies combine keywords across three dimensions: (i) AI/ML and data-driven modeling, (ii) building energy benchmarking (e.g., EUI, load prediction, retrofit targeting), and (iii) occupant-centered outcomes such as thermal comfort, IAQ, and health. Findings are synthesized to identify methodological advances and remaining gaps in integrated benchmarking for climate resilience.

The findings were synthesized into a three-part framework: (i) building energy benchmarking at multiple time scales, (ii) AI-enabled IEQ and health assessment, and (iii) balancing energy savings with occupant comfort. To demonstrate practical application, the chapter incorporates an illustrative case study by [37], which exemplifies how AI-enabled, multi-objective optimization frameworks can operationalize the integration of energy efficiency and occupant comfort discussed in the preceding sections. The case study is used to concretize methodological concepts and illustrate how AI-based benchmarking approaches can be implemented in real-world building operations.

This combined narrative review and illustrative case study approach provides an accessible synthesis of current advances while highlighting emerging opportunities for AI-enabled, occupant-centered benchmarking in support of climate-resilient buildings and cities.

3 AI-Powered Building Performance Benchmarking Approaches

AI-powered benchmarking can be viewed through three scopes: multi-time-scale energy use and performance, IEQ and health, and the trade-off between energy efficiency and comfort. These themes frame how emerging methods extend benchmarking from energy metrics to more holistic, occupant-centered evaluations.

3.1 Building Energy Benchmarking at Multiple Time Scales

As addressed in the Introduction section, building performance benchmarking is a critical approach to trade and potentially reduce energy loads from single to urban levels [43]. Energy benchmarking approaches span three time scales: annual, monthly, and daily/hourly [31]. Depending on the user's purpose, applying high-resolution smart meter data like daily/hourly scales, the benchmarking systems can assess energy performance on a more detailed basis [54], while annual and monthly scales are able to analyze more seasonal trends to capture a bigger picture regarding building energy conservation and decarbonization [2].

On the annual scale, benchmarking frameworks are among the most widely adopted tools for evaluating building energy performance compared to other time levels [45]. Studies across diverse regions, including North America, Europe, Asia, etc., have shown that traditional ML model approaches such as multiple linear regression (MLR) and decision trees (DT), along with more advanced ensemble models like LightGBM (LGBM) and XGBoost (XGB) [47], can provide reliable predictions and generate practical performance benchmarks [6, 14]. Several studies also developed deep learning (DL) or hybrid models to leverage both numeric and imagery data (e.g., GIS and satellite images) to accurately estimate energy usage [52].

Monthly benchmarking provides insights into seasonal variations compared to annual benchmarking, though fewer studies have explored it [27]. Ensemble learning (such as XGB and LGBM) or DL models are among the most reliable approaches tested at this scale in general [34, 48]. Daily and hourly benchmarking offers greater temporal granularity than annual or monthly approaches, and DL methods, specifically Artificial Neural Networks (ANNs) and Long Short-Term Memory (LSTMs), are more effective [54]. These models excel at capturing complex, non-linear energy consumption patterns, particularly when incorporating IEQ factors and occupant behavior [38, 58]. Additionally, daily and hourly benchmarking are more closely associated with occupant comfort and health than larger time scales.

Several notable limitations exist in current research on data-driven energy benchmarking. First, although benchmarking techniques have advanced considerably, many studies have not examined the role of XAI, which can reveal how different input variables influence energy use, offering valuable insights for energy management and conservation [38]. Second, most work continues to rely mainly on tabular datasets, with relatively few efforts incorporating satellite imagery, street-view data, or GIS layers to strengthen contextual understanding and predictive accuracy, despite the promise of this direction [39]. Third, most studies are confined to a single time scale, with only limited work exploring benchmarking across two or more temporal scales, which constrains the understanding of dynamic building performance [2, 30]. Lastly, while some studies combine fundamental building attributes (e.g., area, age, and operational patterns) with IEQ or health factors, the majority overlook occupant-centered information, limiting the ability to link energy loads with occupant health outcomes [19, 20].

3.2 Indoor Environmental Quality (IEQ) and Occupant Health

IEQ, encompassing multiple interrelated factors, including thermal comfort, air quality, lighting, and acoustics [18], plays a pivotal role in determining not only the comfort of building occupants but also their health, productivity, and long-term well-being [3, 51]. People in modern society spend approximately 90% of their time indoors [10, 32, 49], and IEQ significantly affects individual health [9]. Thus, IEQ and occupant health become one pillar, alongside energy efficiency, in building science research. A systematic, quantitative approach can be provided through benchmarking buildings to assess IEQ, enabling a comprehensive evaluation of their performance relative to established standards and best practices [16].

Unlike conventional benchmarking systems that rely on static thresholds, AI-powered approaches can dynamically account for contextual and occupant-specific variables [42]. Recent advancements in ML and DL provide new opportunities to overcome these challenges by leveraging large-scale datasets that include weather conditions, occupancy patterns, and sensor-based measurements of thermal comfort, air quality, lighting, and acoustics [7]. Through continuous data-driven analysis, AI systems can identify hidden patterns, predict energy consumption, optimize IEQ parameters, and link environmental conditions with psychological and physiological health outcomes [42]. As a result, AI-enabled benchmarking not only supports energy efficiency and comfort but also establishes a real-time, adaptive framework for assessing how building operations influence occupant well-being.

Despite these advancements, several gaps remain in integrating AI into building performance benchmarking for IEQ and occupant health. First, most existing studies focus on single-dimensional outcomes such as thermal comfort or energy savings. In contrast, holistic approaches that simultaneously address multiple IEQ factors and health outcomes are still limited [33]. Second, although AI models have shown strong predictive capabilities, their interpretability and ability to provide actionable insights for building operators and policymakers require further development [53]. Third, most datasets used for AI-driven benchmarking are derived from controlled experiments or limited case studies, raising concerns about scalability, generalizability, and transferability across diverse building types and climatic contexts [5]. Finally, the direct linkage between environmental variables and robust indicators of physiological and psychological health is underexplored, with most studies relying on proxy measures or self-reported data [22]. Addressing these gaps presents a critical research opportunity to develop explainable, multi-dimensional, and health-centered AI benchmarking frameworks that not only enhance building performance but also actively promote occupant well-being.

3.3 Balance Energy Saving and Indoor Comfort

One of the central challenges in building performance benchmarking is the trade-off between energy efficiency and indoor comfort [23, 26]. Energy-saving strategies, such as lowering ventilation rates, reducing lighting intensity, and expanding acceptable thermal bands, may compromise IEQ and negatively impact indoor comfort. Conversely, prioritizing indoor comfort without constraints often leads to excessive energy consumption, undermining sustainability targets. In the economic domain, the Second-best theory is frequently used to optimize trade-off decisions [41], which can be transferred to balance the needs of energy saving and indoor comfort. Despite a few studies exploring this trade-off, it remains a significant research gap in the building industry [21].

In addition, the interaction between building energy systems and occupant needs is inherently dynamic [44]. For example, reducing HVAC operation during off-peak hours may conserve energy, but it can also cause discomfort due to temperature or humidity fluctuations. Similarly, automated shading systems that minimize cooling loads may reduce daylight access, impacting visual comfort and productivity. Benchmarking studies increasingly emphasize that "efficiency" metrics must be balanced with indicators of comfort, satisfaction, and health [23].

AI-powered multi-objective optimization has shown strong potential in balancing energy and comfort [44]. AI algorithms can process real-time data streams from building sensors, weather forecasts, and occupancy patterns to optimize system performance under multiple constraints [17]. For example, reinforcement learning can learn adaptive HVAC control policies that minimize energy while maintaining comfort thresholds [1, 11]. Predictive models can anticipate peak load events and adjust setpoints preemptively to reduce discomfort during demand response periods [24]. Occupant-centric AI models integrate feedback (e.g., thermal comfort ratings or wearable sensor data) into control strategies, ensuring energy savings without compromising human well-being [36]. With advanced AI techniques, benchmarking becomes comprehensive and equitable. In the future, benchmarking tools will likely integrate adaptive comfort models, digital twins, and personalized comfort prediction to dynamically balance energy and comfort across diverse occupant groups.

4 Case Study

In this section, the chapter illustrates how advanced AI models can be applied to building performance benchmarking across energy efficiency, IEQ, and health outcomes through a case study. [37] developed a multi-objective optimization framework for air-conditioning and mechanical ventilation (ACMV) systems to address the trade-off between thermal comfort and energy consumption. The model employs four decision variables that significantly influence both thermal and energy performance and builds semi-empirical models to represent component power consumption (fan, pump, and compressor) and their dependence on air temperature. Thermal comfort is modeled with inputs including environmental and personal factors. Multiple ANN models are trained to predict both thermal comfort and energy use; the ANNs are then coupled with an optimization algorithm to search for optimal operating settings. In experiments, the model achieves very high goodness-of-fit, and the optimization yields reductions in energy consumption (~16.5%) while improving thermal comfort (~49.1%) compared to baseline operation. The study collectively addresses three central questions: (i) what challenges arise in applying AI to building performance; (ii) how occupant comfort can be enhanced with high-energy efficiency; and (iii) how these techniques are practically implemented.

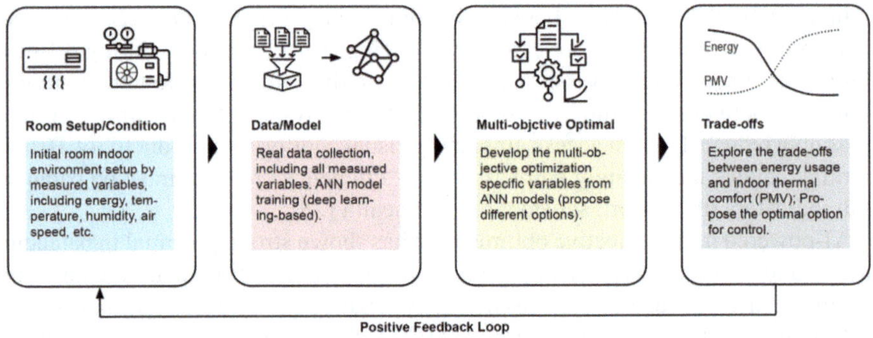

Fig. 2 Workflow of the AI-enabled case study, illustrating how IEQ data and energy consumption are jointly modeled through ANN training and multi-objective optimization to balance energy efficiency and occupant comfort, an essential capability for resilient building operation under climate stress. Summarized from [37], visualized by authors

4.1 Research Approach

The authors collected real-world data from an educational building at National Taiwan University of Science and Technology, including time-series data on indoor environmental parameters and HVAC operational data. The study formulates objective functions using DL-ANN models that quantify both energy use (e.g., HVAC energy load) and occupant thermal comfort (which involves several metrics, such as temperature deviation from comfort setpoints). Then, the decision variables (e.g., thermostat setpoints, HVAC operating schedules, or control parameters) are identified by ANN model predictions that can influence both thermal comfort and energy efficiency.

With these predictive models in place, the study establishes a multi-objective optimization problem that aims to minimize energy consumption while maximizing occupant comfort. They used optimization algorithms to explore trade-offs among these two objectives. Various HVAC control strategy parameters are treated as decision variables—such as setpoints or schedules affecting heating, cooling, and ventilation—that can be adjusted. The optimization identifies candidate strategies and selects the control strategies that best balance energy consumption and thermal comfort. The general workflow of this study is shown in Fig. 2.

4.2 Findings

After real-world data collection and pre-processing, the study employs advanced multiple ANN algorithms to train the models. As illustrated in Fig. 3, five ANN training algorithms are trained and compared. The Levenberg–Marquardt method achieved the lowest MAE and the highest R^2 values among the five ANN models. As

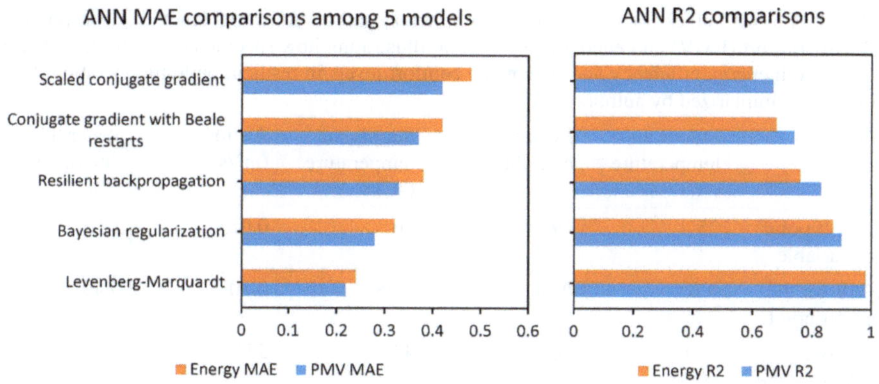

Fig. 3 Performance comparison of five ANN training algorithms using MAE (left) and R^2 (right), demonstrating the predictive reliability required for AI-enabled benchmarking frameworks that simultaneously evaluate energy consumption and occupant comfort in resilient building operations. Data from [37], visualized by authors

such, it is selected to train the Predicted Mean Vote (PMV) and energy consumption (EC) datasets.

The study identified optimal cut-off points for key indoor environmental variables to balance energy efficiency with thermal comfort, measured by the PMV. As shown in Table 1, the cut-off conditions include an indoor air temperature of 25 °C, a relative humidity of 49%, a supply air temperature of 16 °C, an airflow rate of 0.055 m^3/s, and a water flow rate of 2.75 m^3/s. At these thresholds, PMV values remained close to neutral (ranging from 0.20 to 0.40), indicating acceptable comfort conditions. Corresponding energy use varied across parameters, from 300 kWh (relative humidity) to 625 kWh (water flow rate). These results highlight the trade-offs inherent in HVAC control, showing how specific setpoints can simultaneously minimize energy consumption while maintaining comfort within acceptable PMV ranges.

Figure 4 summarizes the optimal control settings together with the resulting PMV and EC. The base case represents the room's reference operating condition. According to ISO 7730, both the base case and solution case C meet Class C, which represents the minimum acceptable level of thermal comfort. Case A attains Class A, the highest comfort category, while case B falls into Class B, an intermediate comfort level. Case B demonstrates a significant improvement over the baseline, indicating lower energy use and higher thermal satisfaction. Additionally, although solution case A has the highest comfort level, it consumes significantly more energy (~165.1% more than the base case).

Table. 1 Cut-off values for key indoor environmental and HVAC variables with corresponding thermal comfort (PMV) and energy consumption, illustrating how specific control thresholds can support both energy efficiency and acceptable comfort levels in resilient building operation. Data from [37], summarized by authors

Variable	Indoor air temperature (°C)	Relative humidity (%)	Supply air temperature (°C)	Airflow rate (m^3/s)	Water flow rate (m^3/s)
Cut-off point for variable	25	49	16	0.055	2.75
PMV for the cut-off point	0.20	0.24	0.28	0.20	0.40
Energy use (kWh) for the cut-off point	400	300	450	322	625

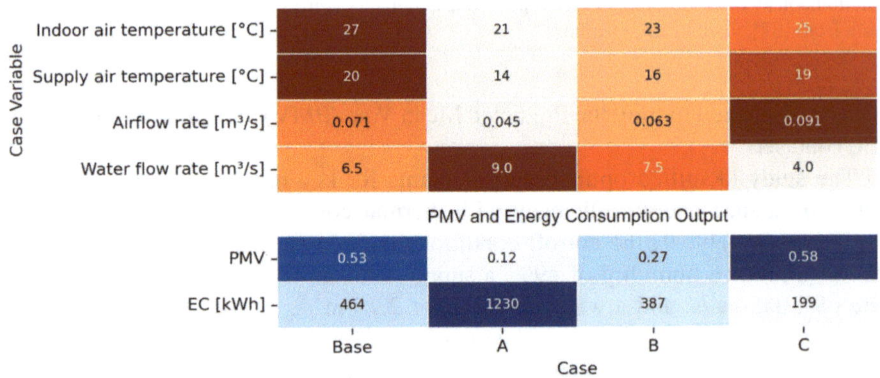

Fig. 4 Multi-objective optimization results illustrating trade-offs between energy consumption and thermal comfort (PMV) across different control strategies. The figure demonstrates how the model can identify operating points that improve occupant comfort while reducing energy use, thereby supporting health-protective, resilient building performance. Data from [37], visualized by authors

5 Discussion and Implications

The integration of AI into high-performance building benchmarking brings both opportunities and challenges that extend beyond technical optimization. The chapter highlights three implications. First, AI enables benchmarking across multiple time scales and performance dimensions. By transitioning from annual and monthly total energy/EUI to more granular, adaptive indicators, such as overheating risk, thermal autonomy, and passive survivability, benchmarking can better capture how buildings perform under future climate change scenarios. This shift allows practitioners and policymakers to evaluate not only efficiency but also resilience.

Second, when aggregated across building portfolios, AI-enabled benchmarking can inform city-scale resilience planning and policy. Portfolio-level energy, comfort, and health indicators support district energy planning, urban heat mitigation, and resilience-oriented building codes. Spatial benchmarking patterns also reveal inequities in thermal autonomy and indoor air quality, identifying neighborhoods most vulnerable during extreme heat or pollution events. By making these disparities visible, AI-driven benchmarking guides targeted retrofits and policy interventions that advance both community resilience and environmental equity.

Third, AI broadens the scope of benchmarking to encompass occupant health and equity. Traditional approaches emphasize carbon reduction and energy metrics, while often overlooking the social and physiological outcomes of building operations. As climate change increases exposure to wildfire smoke, heatwaves, and power outages, AI-based frameworks can dynamically link IEQ factors with health and productivity outcomes. Embedding equity considerations into benchmarking ensures resilience planning.

Lastly, implementation requires balancing predictive sophistication with interpretability and practical deployment. While DL and hybrid models deliver higher performance, their "black box" nature can hinder adoption by building managers and policymakers. XAI, human-in-the-loop controls, and digital twins are crucial for translating predictions into actionable decisions. Additionally, data limitations—such as fragmented utility records, sparse sensor deployments, and limited health-linked datasets—need to be addressed to ensure scalability across building types and regions.

Collectively, these insights suggest that benchmarking must shift from single-objective, energy-focused checks to multi-objective, adaptive, occupant-centered evaluations. Practitioners should integrate AI into building management systems for real-time optimization; researchers should advance interdisciplinary methods that link building science, public health, and data science; and policymakers should adopt standards that embed resilience and equity alongside efficiency. Framing AI as both a predictive engine and a vehicle for inclusive, climate-ready design can turn benchmarking into a catalyst for a healthier, more equitable, and sustainable built environment.

6 Conclusion

High-performance buildings can no longer be evaluated solely through the lens of energy efficiency or carbon reduction. As climate change accelerates the frequency and severity of extreme weather events, the resilience of buildings should also be judged by how well they protect human comfort and health. This chapter has demonstrated that AI-enabled benchmarking frameworks provide a new pathway toward integrated performance assessment, capturing operational metrics, such as EUI, as well as adaptive indicators, including thermal autonomy, passive survivability, and real-time indoor environmental quality. By leveraging ML and advanced AI, these approaches allow for dynamic, scalable, and inclusive evaluations across diverse building types and scales. Additionally, integrating health and equity into benchmarking frameworks underscores that resilience must extend beyond technical efficiency to social inclusivity. Communities with fewer resources often bear the heaviest burden from power outages, poor air quality, and heat stress. AI-powered tools that combine building data with social and health metrics can help ensure that resilience strategies are not only technologically advanced but also equitable and human-centered.

Looking ahead, the convergence of data-driven methods, occupant-centric design, and emerging AI technologies signals a paradigm shift in how we conceive of high-performance buildings. Rather than optimizing energy and comfort in isolation, future benchmarking frameworks will balance multiple objectives in real time, adapt to evolving climates, and promote healthier, more sustainable environments. In this way, AI-enabled benchmarking is positioned not merely as a technical optimization tool but as a catalyst for transforming buildings into adaptive, health-supporting nodes within a resilient urban system. By enabling proactive, data-driven, and equitable decision-making, these frameworks can help cities manage climate risks more effectively and prepare the built environment for an increasingly uncertain future.

Acknowledgements The authors would like to express great appreciation for the funding support from the College of Architecture at the University of Nebraska-Lincoln and the University of Nebraska Foundation.

References

1. Al Sayed K, Boodi A, Broujeny RS, Beddiar K (2024) Reinforcement learning for HVAC control in intelligent buildings: A technical and conceptual review. Journal of Building Engineering 95:110085
2. Andrews A, Jain RK (2023) Modeling the Decarbonization potential of a time-of-use building energy benchmarking model at the urban scale. In computing in civil engineering (pp. 304–312)
3. Arif M, Katafygiotou M, Mazroei A, Kaushik A, Elsarrag E (2016) Impact of indoor environmental quality on occupant well-being and comfort: A review of the literature. Int J Sustain Built Environ 5(1):1–11
4. Amasyali K, El-Gohary NM (2018) A review of data-driven building energy consumption prediction studies. Renew Sustain Energy Rev 81:1192–1205
5. Andersen KH, Johra H, Schaffer M, Marszal-Pomianowska A, Knudsen HN, Heiselberg PK, O'Brien W (2024) Exploring occupant detection model generalizability for residential buildings using supervised learning with IEQ sensors. Build Environ 254:111319
6. Arjunan P, Poolla K, Miller C (2020) EnergyStar++: Towards more accurate and explanatory building energy benchmarking. Appl Energy 276:115413
7. Avci AB (2025) Machine learning-based prediction of thermal comfort: exploring building types, climate, ventilation strategies, and seasonal variations. Building Research & Information: 1–18.
8. Awadh O (2017) Sustainability and green building rating systems: LEED, BREEAM, GSAS and Estidama critical analysis. Journal of Building Engineering 11:25–29
9. Bangwal D, Kumar R, Suyal J, Ghouri AM (2023) Does AI-technology-based indoor environmental quality impact occupants' psychological, physiological health, and productivity?. Annals of Operations Research: 1–19.

10. Bie H, Guo B, Li T, Loftness V (2025) A review of air-side economizers in human-centered commercial buildings: Control logic, energy savings, IAQ potential, faults, and performance enhancement, Energy and Buildings: 115389.
11. Boutahri Y, Tilioua A (2025) Reinforcement learning for HVAC control and energy efficiency in residential buildings with BOPTEST simulations and real-case validation. Discover Computing 28(1):45
12. Capeluto G, Ochoa CE (2016) Intelligent envelopes for high-performance buildings: Design and strategy. Springer.
13. Chen X, Yang H, Wang Y (2017) Parametric study of passive design strategies for high-rise residential buildings in hot and humid climates: miscellaneous impact factors. Renew Sustain Energy Rev 69:442–460
14. Chen Y, Tan H, Berardi U (2018) A data-driven approach for building energy benchmarking using the Lorenz curve. Energy and Buildings 169:319–331
15. Chen Y, Ye Y, Liu J, Zhang L, Li W, Mohtaram S (2023) Machine learning approach to predict building thermal load considering feature variable dimensions: an office building case study. Buildings 13(2):312
16. Dam-Krogh EP, Clausen G, Toftum J (2024) Optimising indoor environmental quality: Development of a benchmarking tool for valuing performance in office buildings. Build Environ 266:112120
17. Das HP, Lin YW, Agwan U, Spangher L, Devonport A, Yang Y, Spanos CJ (2024) Machine learning for smart and energy-efficient buildings. Environmental Data Science 3:e1
18. Deng Z, Dong B, Guo X (2024) Zhang J (2024) Impact of Indoor Air Quality and Multi-domain Factors on Human Productivity and Physiological Responses: A Comprehensive Review. Indoor Air 1:5584960
19. Geraldi MS, Ghisi E (2020) Mapping the energy usage in Brazilian public schools. Energy and Buildings 224:110209
20. Geraldi MS, Ghisi E (2022) Integrating evidence-based thermal satisfaction in energy benchmarking: A data-driven approach for a whole-building evaluation. Energy 244:123161
21. Dong Z, Luo X, Zhao K, Ge J, Chan IY (2025) Comprehensive assessment method for building environmental performance: Trade-off between indoor environmental quality and life cycle carbon emissions. Build Environ 272:112598
22. Grasso-Cladera A, Arenas-Perez M, Wegertseder-Martinez P, Vilina E, Mattoli-Sanchez J, Parada FJ (2025) Neuroscientific Insights into the Built Environment: A Systematic Review of Empirical Research on Indoor Environmental Quality, Physiological Dynamics, and Psychological Well-Being in Real-Life Contexts. Int J Environ Res Public Health 22(6):824
23. Guillén EE, Samuelson HW, Laurent JGC (2019) Comparing energy and comfort metrics for building benchmarking. Energy and Buildings 205:109539
24. Hao J (2020) Deep reinforcement learning for the optimization of building energy control and management (Doctoral dissertation, University of Denver)
25. IEA (2021) Net zero by 2050. IEA, Paris. https://www.iea.org/reports/net-zero-by-2050
26. Jakob M (2007) The Trade-offs and Synergies Between Energy Efficiency, Costs and Comfort in Office Buildings. In Proc. of CISBAT.
27. Jeun W, Hwang J, Suh D, Otto MO (2025) Data-driven approach for electricity consumption benchmarking of multi residential buildings. Energy and Buildings 345:115944
28. Kontokosta C (2011) Greening the regulatory landscape: The spatial and temporal diffusion of green building policies in US cities. Journal of Sustainable Real Estate 3(1):68–90
29. Kontokosta CE, Tull C (2017) A data-driven predictive model of city-scale energy use in buildings. Appl Energy 197:303–317
30. Lamagna M, Nastasi B, Groppi D, Nezhad MM, Garcia DA (2020) Hourly energy profile determination technique from monthly energy bills. In Building Simulation, Vol 13, No 6. Beijing, Tsinghua University Press, pp 1235–1248.
31. Lee J, Kim TW, Koo C (2022) A novel process model for developing a scalable room-level energy benchmark using real-time bigdata: Focused on identifying representative energy usage patterns. Renew Sustain Energy Rev 169:112944

32. Lee MJ (2024) A Novel approach to indoor environment assessment: Artificial Intelligence of Things (AIoT) framework for improving occupant comfort and health in educational facilities (Doctoral dissertation, Virginia Polytechnic Institute and State University)
33. Leccese F, Rocca M, Salvadori G, Belloni E, Buratti C (2021) Towards a holistic approach to indoor environmental quality assessment: Weighting schemes to combine effects of multiple environmental factors. Energy and Buildings 245:111056
34. Li T, Bie H, Lu Y, Sawyer AO, Loftness V (2024) MEBA: AI-powered precise building monthly energy benchmarking approach. Appl Energy 359:122716
35. Li T, Liu T, Sawyer AO, Tang P, Loftness V, Lu Y, Xie J (2024) Generalized building energy and carbon emissions benchmarking with post-prediction analysis. Developments in the Built Environment 17:100320
36. Liang X, de Nijs F, Say B, Wang H (2025) Human-in-the-loop Ai for HVAC management enhancing comfort and energy efficiency. In Proceedings of the 16th ACM International Conference on Future and Sustainable Energy Systems, pp 359–370.
37. Lin CJ, Wang KJ, Dagne TB, Woldegiorgis BH (2022) Balancing thermal comfort and energy conservation–A multi-objective optimization model for controlling air-condition and mechanical ventilation systems. Build Environ 219:109237
38. Liu C, Li Y, Chen H, Xing L, Zhang S (2024) Energy-saving potential benchmarking method of office buildings based on probabilistic forecast. Journal of Building Engineering 95:110282
39. Ma J, Cheng JC (2016) Estimation of the building energy use intensity in the urban scale by integrating GIS and big data technology. Appl Energy 183:182–192
40. Mansoor S, Farooq I, Kachroo MM, Mahmoud AED, Fawzy M, Popescu SM, Ahmad P (2022) Elevation in wildfire frequencies with respect to the climate change. J Environ Manage 301:113769
41. Markovits RS (1997) Second-Best Theory and Law & (and) Economics: An Introduction. Chi-Kent L Rev 73: 3
42. Mehmood MU, Chun D, Han H, Jeon G, Chen K (2019) A review of the applications of artificial intelligence and big data to buildings for energy-efficiency and a comfortable indoor living environment. Energy and buildings 202:109383
43. Na W, Liu S (2024) Benchmarking building energy consumption for space heating using an empirical Bayesian approach with urban-scale energy model. Energy and Buildings 320:114581
44. Ogundiran J, Asadi E, Gameiro da Silva M (2024) A systematic review on the use of AI for energy efficiency and indoor environmental quality in buildings. Sustainability 16(9):3627
45. Papadopoulos S, Kontokosta CE (2019) Grading buildings on energy performance using city benchmarking data. Appl Energy 233:244–253
46. Parrado-Hernando G, Herc L, Pfeifer A, Capellán-Perez I, Bjelić IB, Duić N, Gjorgievski VZ (2022) Capturing features of hourly-resolution energy models through statistical annual indicators. Renewable Energy 197:1192–1223
47. Robinson C, Dilkina B, Hubbs J, Zhang W, Guhathakurta S, Brown MA, Pendyala RM (2017) Machine learning approaches for estimating commercial building energy consumption. Appl Energy 208:889–904
48. Rosenfelder M, Wussow M, Gust G, Cremades R, Neumann D (2021) Predicting residential electricity consumption using aerial and street view images. Appl Energy 301:117407
49. Samet JM, Spengler JD (2003) Indoor environments and health: moving into the 21st century. Am J Public Health 93(9):1489–1493
50. Sartor D, Piette MA, Tschudi W, Fok S (2000) Strategies for energy benchmarking in cleanrooms and laboratory-type facilities, Lawrence Berkeley National Lab (LBNL), Berkeley, CA (United States). https://www.osti.gov/biblio/787084
51. Singh A, Syal M, Grady SC, Korkmaz S (2010) Effects of green buildings on employee health and productivity. Am J Public Health 100(9):1665–1668
52. Streltsov A, Malof JM, Huang B, Bradbury K (2020) Estimating residential building energy consumption using overhead imagery. Appl Energy 280:116018

53. Teixeira B, Carvalhais L, Pinto T, Vale Z (2025) Explainable AI framework for reliable and transparent automated energy management in buildings. Energy and Buildings: 116246
54. Wang Z, Hong T, Piette MA (2020) Building thermal load prediction through shallow machine learning and deep learning. Appl Energy 263:114683
55. Ye Y, Zuo W, Wang G (2019) A comprehensive review of energy-related data for us commercial buildings. Energy and Buildings 186:126–137
56. Zheng H, Zhang D, Schweiker M, Loomans M (2025) A review of building technology solutions and their influence on indoor environmental quality in the healthy building movement. J Build Eng: 112086
57. Zhang L, Wen J, Li Y, Chen J, Ye Y, Fu Y, Livingood W (2021) A review of machine learning in building load prediction. Appl Energy 285:116452
58. Zhou X, Mei Y, Liang L, Fan Z, Yan J, Pan D (2021) A dynamic energy benchmarking methodology on room level for energy performance evaluation. Journal of Building Engineering 42:102837

[illegible]

Integrating Source-Sink Theory and Road Pavement Solar Collectors for Urban Climate Resilience

Hadis Allahyari

1 Introduction

Cities around the world are experiencing rapid urbanization, characterized by horizontal and vertical expansion to meet the needs of a growing population [1]. This urbanization changes land use and land cover (LULC), increases the absorption of solar radiation mainly due to the increase of artificial dark surfaces and building materials characterized by low albedo, higher heat capacity and conductivity compared to non-urban areas, and therefore leads to an increase in temperatures in cities, creating a phenomenon called urban heat island (UHI), where temperatures near the urban air surface are higher than in the rural environment, especially at night [2–6]. Most urban areas are affected by UHI due to urbanization [7]. Several studies have shown that increased energy and water consumption is associated with the UHI effect. During the summer season, UHI contributes significantly to increased energy demand, especially for air conditioning [8]. In addition to the increase in electricity consumption and temperature, UHI also leads to increased pollution intensity in urban areas and affects regional climatology by changing regional wind patterns, increasing humidity, fog and cloud formation and altering precipitation frequency [9]. Warmer air also affects the rate of smog (ozone) formation, which is a major environmental and health problem. These diseases cause many deaths annually as one of the effects of UHI.

As a result, the pattern and magnitude of UHI effects are the subject of numerous studies in urban climatology [10]. Heatwaves, which are characterized by prolonged periods of extremely high temperatures, pose a major challenge for the built environment in cities in several respects [11]. During climate change, the frequency and intensity of heatwaves are increasing. Heatwaves have a synergistic effect with the UHI phenomenon, exacerbating the intensity of urban overheating during heatwaves by 5–10 °C worldwide [12]. Land use planning and heat storage materials in built-up areas

H. Allahyari (✉)
Department of Physics and Astronomy "Augusto Righi," University of Bologna, Bologna, Italy
e-mail: hadis.allahyari2@unibo.it

A. Cheshmehzangi et al. (eds.), *Design Innovation for Climate-Resilient Cities and Built Environments*, Urban Sustainability, https://doi.org/10.1007/978-981-92-2297-1_9

have a major influence on the extent of the UHI phenomenon and thus on the potential impact of heatwaves in cities. As cities become more urbanized, the combined effects of heatwaves and UHI on the built environment may intensify [13]. Given the critical role of the built environment in causing these impacts, understanding the synergistic effects of heatwaves and UHI on public health, building energy consumption and infrastructure is crucial for addressing the associated challenges and developing effective strategies to mitigate them [14].

Traditional approaches to UHI management, such as planting trees, installing cool roofs and increasing green spaces, have been implemented on a large scale. However, these methods have several limitations. In densely built-up environments, the space available for implementing cooling strategies is often limited, which reduces their overall impact [15]. On the other hand, green infrastructure requires constant maintenance, including irrigation and care, which can be resource-intensive and costly. In addition, the benefits of traditional UHI mitigation strategies are not necessarily evenly distributed, so vulnerable communities may be neglected. Another reason is that traditional methods may not be effectively integrated into broader urban development planning and strategy, limiting their long-term sustainability.

These limitations highlight the need for innovative and integrated approaches to UHI management that go beyond traditional methods. Overcoming these challenges requires innovative approaches that combine the principles of source-sink theory with novel urban infrastructure solutions such as solar panels on road surfaces. In studies related to UHI, heat source and heat sink landscapes are defined as landscape types that favor or hinder the development of UHI. In this case, the source-sink theory considers urban areas as dynamic systems in which built-up areas act as "heat sources," while green spaces, water bodies and other cooling elements act as "sinks" that absorb and dissipate heat energy [16].

By strategically increasing sinks and converting specific urban surfaces into energy-absorbing or heat-reducing elements, the intensity of UHI can be reduced while simultaneously generating renewable energy. Solar roadway collectors, which convert solar radiation absorbed by streets into usable thermal or electrical energy, are a promising example of this dual approach. Integrating such technologies into urban planning can create resilient cities that not only reduce ambient temperatures but also contribute to sustainable energy production. Furthermore, combining source-sink optimization with smart placement of solar collector pavements can reduce localized overheating, improve thermal comfort, and reduce energy consumption for cooling, especially during heat waves. This integrated strategy provides a forward-looking framework for urban climate resilience that links urban design, energy management, and environmental sustainability in a holistic manner. Numerous scientific publications have discussed various techniques and approaches for measuring the urban heat island. Each type of urban heat island corresponds to a specific observation method, which can be divided into three main categories: In-situ measurements, remote sensing methods and modeling approaches. In-situ analysis is based on meteorological data recorded by fixed towers or automatic weather stations installed at a height of 2 metres above the ground and distributed in different urban and rural areas. The in—situ technique also includes mobile measurements, which require some pre—and post-processing to ensure the homogeneity of the data.

2 Background Knowledge

2.1 *Urban Heat Island Studies: Measurement and Characterization*

The urban heat island (UHI) phenomenon has been studied in detail to understand how urbanization affects the local microclimate. As mentioned above, UHI refers to the high temperatures observed in urban areas compared to the surrounding rural areas, largely due to the replacement of natural surfaces with impermeable materials such as asphalt and concrete, which have a low albedo and high heat capacity. Surface, canopy, and boundary layer urban heat islands are the three different forms of urban heat islands that can be identified [17]. Numerous scientific publications have discussed various techniques and approaches for measuring the urban heat island. Each type of urban heat island corresponds to a specific observation method, which can be divided into three main categories: In-situ measurements, remote sensing methods and modeling approaches. In-situ analysis is based on meteorological data recorded by fixed towers or automatic weather stations installed at a height of 2 meters above the ground and distributed in different urban and rural areas. The in-situ technique also includes mobile measurements, which require some pre—and post-processing to ensure the homogeneity of the data [18, 19].

Protecting the instrument from the heat generated by the vehicle's exhaust and fitting it with a sensor that has adequate ventilation are, in fact, crucial. Furthermore, by selecting the start and endpoints as a common location, data gathered via the automobile journey must be adjusted for temperature variations that occur during the traverse. Data obtained from satellite sensors forms the basis of the second technique, remote sensing analysis. The literature has also covered other methods that enable the retrieval of urban heat islands based on this second approach, such as the use of data gathered by aircraft-mounted temperature sensors or tethered balloons/airships flown across an urban region at a certain height from the surface [20–24]. Regarding the last approach, it consists of proceeding by computer modeling.

2.2 *Source-Sink Theory in Urban Climatology*

Some researchers have incorporated the source-sink theory into urban climate studies. Originally applied in air pollution and carbon cycle research, the source-sink theory has expanded to cover various ecological fields. In ecology, source landscapes are those that facilitate ecological processes, while sink landscapes are those that hinder their development. In studies related to UHI, heat source and sink landscapes are defined as landscape types that facilitate or hinder the development of UHI, respectively [25]. Researchers have identified various LULCs as heat source and sink landscapes and have examined the spatial characteristics and factors affecting these landscapes. The application of source-sink theory to UHI is highly appropriate because UHI occurs due to the uneven spatial distribution of heat source and sink landscapes in cities or the excessive concentration of heat source landscapes. Effective application of source-sink theory can reduce UHI by spatially allocating different landscape types, especially in urban environments where land resources are scarce [26] (Fig. 1).

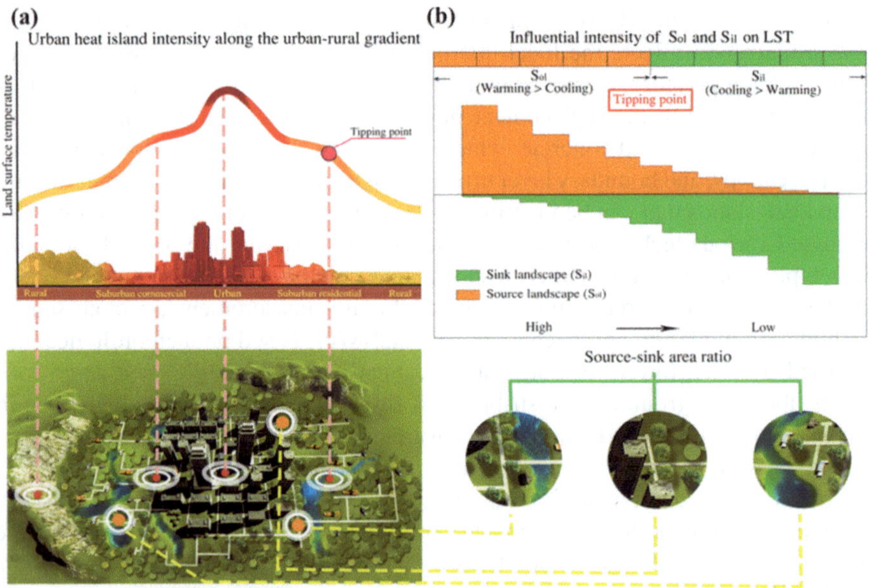

Fig. 1 Urban heat island (UHI) effect schematic diagram. an urban-rural gradient in UHI intensity. This subgraph shows a tipping point where the combined impacts of source and sink landscapes on UHI dramatically diminished when the urban–rural gradient reached a certain range from the city center. b Land surface temperature (LST) is affected by the intensity of the heat source and sink area under various heat source-sink area ratios. In this subgraph, source area is the dominant feature when the ratio of source area exceeds the tipping point; conversely, when the ratio of source area exceeds the tipping point, the warming influence of the source area is greater than the cooling effect of the sink area

2.3 Road Pavement Solar Collectors: Concepts and Categories

Road Pavement Solar Collector (RPSC) systems are cutting-edge technologies that convert pavement and road surfaces from passive solar radiation absorbers into infrastructures that actively gather energy. These systems can capture, storing, and reusing solar energy for thermal and electrical applications by incorporating collectors into or beneath the pavement. A variety of configurations are available, such as high-conductivity surfacing with shallow heat exchangers intended to improve thermal transfer, photovoltaic or thermoelectric pavements that directly convert a portion of surface irradiance into electricity, and thermal collectors with embedded pipes or conductive layers that circulate a working fluid to transfer and store heat (e.g., for district heating or seasonal storage) [27–29]. In addition to providing co-benefits including lower building cooling loads and local energy supply for ventilation or pumping, RPSC systems also help mitigate urban heat islands by reducing daytime surface temperatures. Mechanical durability, skid resistance, maintenance costs, and integration with utility and drainage systems are still issues, though, particularly when traffic volumes are high [29, 30]. All things considered, RPSC technology integration into urban planning offers a dual approach of climate resilience and sustainable energy generation, connecting energy management and urban cooling. As a result, the types of collectors, technological designs, and end-use applications all influence RPSC systems. These systems fall into the following categories:

1. Pavement with integrated thermal collectors: These are made up of highly conductive layers or a system of pipes positioned beneath the road surface's wearing course. The system transfers heat that has been absorbed to a fluid that is in circulation, usually a mixture of water and glycol. In Europe, especially in the Netherlands and Switzerland, where pilot projects have shown seasonal storage in aquifer thermal energy storage (ATES) or borehole thermal energy systems (BTES), such designs are well-established. For greater energy usage, heat can be provided to heat pumps or stored underground for use in space heating in the winter [31, 32].
2. Integrated thermoelectric and photovoltaic pavement: Another type of RPSC is the integration of solar and thermoelectric materials directly into the pavement surfaces. These materials can be semi-transparent photovoltaic (PV) modules, thin-film solar coatings, or thermoelectric generators that absorb sunlight and convert part of it into usable electricity. These systems are particularly useful for local energy needs. For example, systems for powering streetlights, traffic sensors, or other small-scale urban applications close to the road are of this type. The use of these systems on a large scale and on high-traffic roads still faces limitations. The main challenges in maintaining these systems are ensuring durability under constant vehicle loads, maintaining sufficient resistance and safety against slipping, and addressing the relatively high installation and maintenance costs [33]. Research is exploring solutions such as advanced encapsulation techniques and flexible PV layers that can better withstand mechanical stress, heavy traffic,

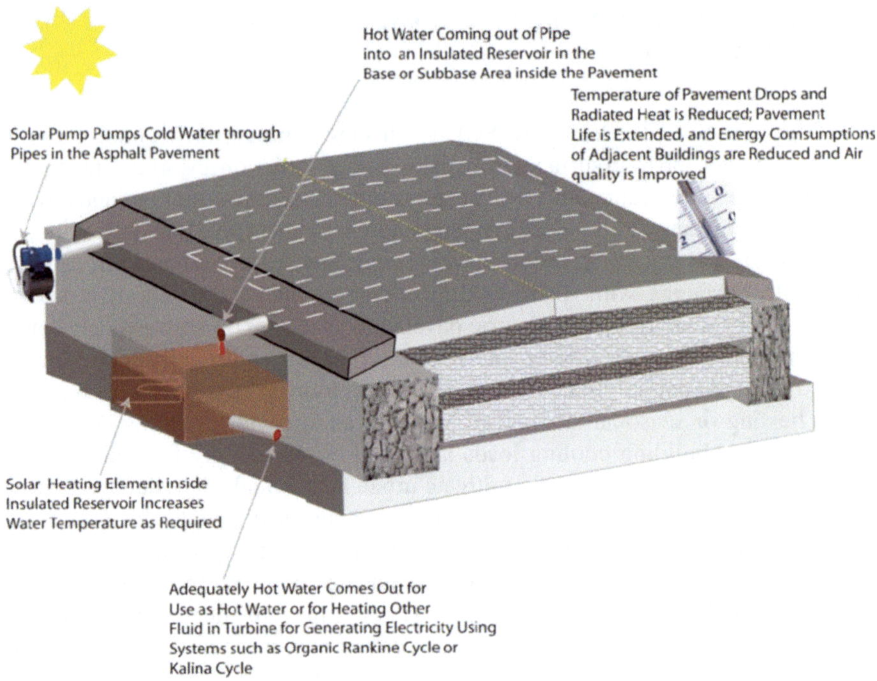

Fig. 2 Concept of harvesting energy from pavements and reducing pavement temperature [28]

and thermal expansion, paving the way for more robust applications in the future [34].

3. High-conductivity surfaces with heat exchangers

   In this case, engineered concretes with additives such as graphite, steel fibers, or phase change materials are used to improve thermal conductivity. These materials increase heat transfer to shallow subsurface collectors. These pavements, combined with embedded heat exchangers, hold promise for urban areas with high cooling demands, as they can reduce peak surface temperatures while providing low-grade heat to regional energy networks [35] (Fig. 2).

3 Methods: Modeling, Design and Assessment Workflow

An outline of the Methodological Process.

Section 3 presents an integrated modeling, design, and assessment process developed to identify urban heat sources, prioritize intervention areas, and assess the potential impact of installing road pavement solar collectors (RPSC). The methodology follows a coherent three-step planning process (Fig. 3):

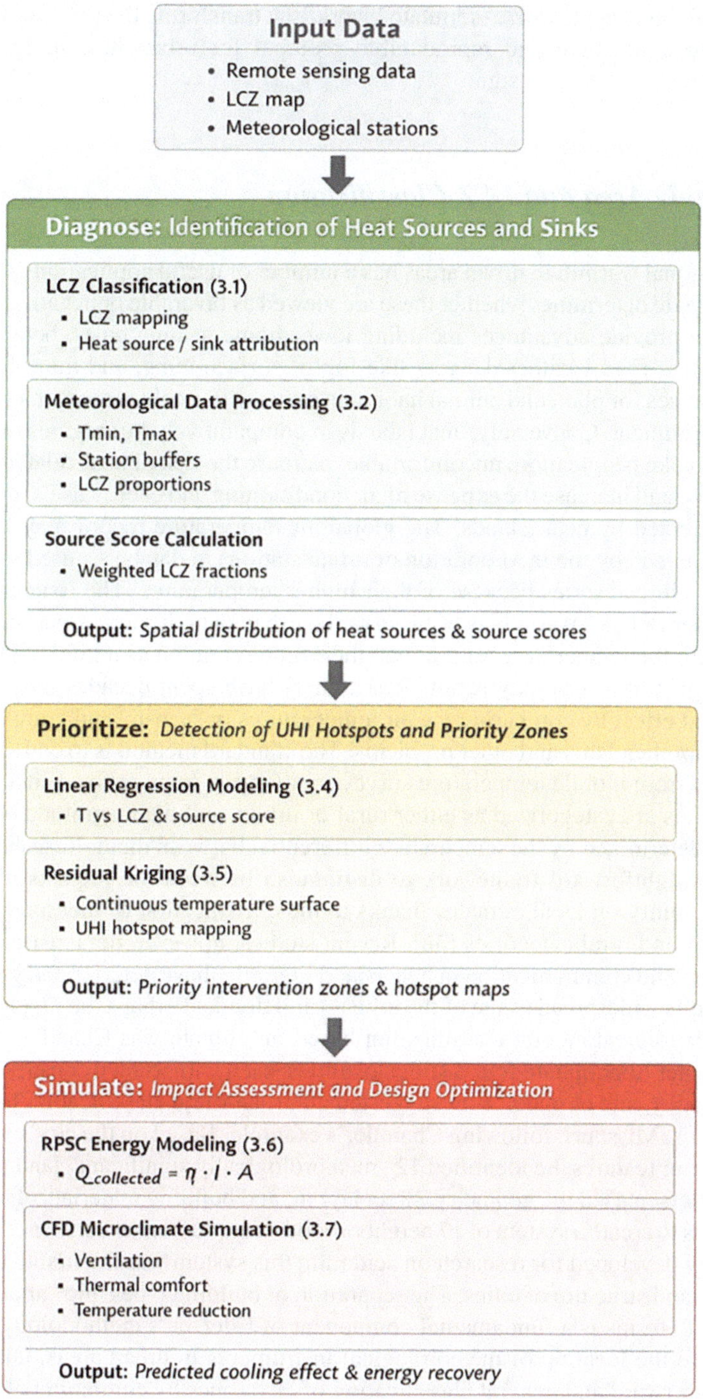

Fig. 3 Methodological Workflow for RPSC-Based Urban Heat Mitigation

This "diagnose–prioritize–simulate" workflow transforms the proposed methodology into a practical and reproducible protocol for urban heat mitigation and climate-sensitive urban design.

3.1 Study Area and LCZ Classification

The additional warmth in urban areas has a number of useful applications. The city's macroclimate determines whether these are viewed as favorable or negative. The heat island can provide advantages including lower home heating costs, better outdoor comfort, less road weather dangers like fog or surface icing, and more hospitable circumstances for plant and animal habitat in cities with a cold season or a somewhat cold environment. Conversely, heat islands in comparatively hot regions or times of year can make people more uncomfortable, increase the risk of heat-related illnesses and deaths, and increase the expense of air conditioning and energy use. Climatology is also affected by heat islands. The global air temperature record may have been "contaminated" by the incorporation of urban stations in databases used to evaluate historical climate series because of their higher temperatures. The issue is whether the time series has a warm bias because of the urban data. If more urban stations are included in the temperature record than their representation as a kind of land cover on Earth justifies, bias may result. Researchers have spent decades measuring the heat island effect by comparing the air temperatures in "urban" and "rural" areas to fully comprehend this and other problems. The standard method is to collect temperature data from mobile temperature surveys and/or two or more fixed sites at screen height. Sites are categorized as either rural or urban, and the magnitude of the heat island is determined by the temperature differential between them. Researchers now have a straightforward framework to distinguish between the impacts of the city and the country on local climates thanks to the classification of measurement sites into urban and rural categories [36]. Recent studies, however, demonstrate that the techniques and communication in heat island literature have deteriorated greatly as a result of this widespread usage of the urban–rural divide. Perhaps the first heat island researcher to create a city classification based on climate was Chandler. He separated Greater London into four distinct localities, each with its own constructed form, physiography, and climate. Auer suggested an urban–rural classification for the city of St. Louis, Missouri, following Chandler's example. Based on the city's vegetation and building features, he identified 12 "meteorologically significant" land uses in St. Louis. Ellefsen used the geometry, street layout, and building materials of ten American towns to create a system of 17 neighborhood-scale "urban terrain zones" (UTZs). Originally developed for research on acid rain, this system was the first to depict the materials and structure of cities. The separation of building types into "attached" and "detached" forms is a fundamental component of Ellefsen's methodology. In order to improve the location of meteorological instruments in urban areas, Oke created a straightforward and general classification of city zones by combining elements of

Auer's and Ellefsen's systems. According to his plan, the city's geography is separated into seven uniform areas known as "urban climate zones" (UCZs), which vary from heavily urbanized areas to semi-rural areas. The urban structure (building/street proportions), cover (permeability), fabric (materials), metabolism (human activity), and tendency to alter the natural, or "preurban," surface climate are the characteristics that set the zones apart. Most recently, "urban zones for characterizing energy partitioning," or UZEs, were created by Loridan and Grimmond. Threshold levels for the constructed surface and active vegetative fractions of cities—where "active" refers to energy exchange—define their classes. Continuing from this framework, Local Climate Zones (LCZs) extend these classifications by providing a more detailed spatial categorization of urban and non-urban areas based on surface cover, building geometry, and vegetation. The types of LCZ are given on the table [37] (Table 1).

Each LCZ can be associated with its role in the urban heat island (UHI) effect, acting either as a heat source or a heat sink. Densely built-up and high-rise zones (LCZ 1–3) with large areas of impervious surfaces and minimal vegetation tend to function as strong heat sources, storing and radiating heat into the surrounding air. Medium-density or open low-rise zones (LCZ 4–6) generate moderate heat, partially mitigated by permeable surfaces and scattered greenery. Industrial and commercial zones (LCZ 7) often contribute additional heat due to dark surfaces and anthropogenic activities, while sparsely built, vegetated, or water-covered zones (LCZ 8–10) act as heat sinks, promoting cooling through shading, evapotranspiration, and surface permeability. By mapping these zones and their relative thermal roles, LCZs provide a practical framework for quantifying spatial variations in UHI intensity and supporting urban planning strategies aimed at mitigating excessive heat in cities. Bologna was chosen as the study area due to its diverse urban morphology, which includes high-density historical cores, medium-rise residential neighborhoods, and extensive green spaces on the city's periphery, making it an ideal case for analyzing the spatial variability of urban heat sources and sinks using the LCZ classification. Accordingly, I mapped the Bologna LCZ to visualize the spatial distribution of urban heat sources and sinks across the city (Fig. 4).

Accordingly, the role of each LCZ in increasing or decreasing heat in the city of Bologna can be summarized as follows (Table 2).

This spatial framework acts as a foundation that links urban morphology and thermal characteristics to enable the use of a combination of source-sink theory and RPSC systems. Beyond its role as a site-specific case study, Bologna functions here as a test bed for extracting transferable principles for RPSC-based urban heat mitigation. The results indicate that the highest-priority candidates for RPSC installation are not uniformly all dense urban areas, but rather specific functional urban typologies characterized by a combination of high impervious surface fraction, strong solar exposure, and sufficient pavement continuity. In particular, compact mid-rise and low-rise commercial strips (LCZ 2–3) and large low-rise industrial and commercial zones (LCZ 8) consistently emerged as the most favorable environments for RPSC deployment, due to their strong heat storage capacity, extensive paved surfaces, and limited shading constraints. Wide arterial roads crossing these zones further

Table 1 LCZ Classification [37]

Built types	Definition	Land cover types	Definition
1. Compact high-rise 	Dense mix of tall buildings to tens of stories. Few or no trees. Land cover mostly paved. Concrete, steel, stone, and glass construction materials.	A. Dense trees 	Heavily wooded landscape of deciduous and/or evergreen trees. Land cover mostly pervious (low plants). Zone function is natural forest, tree cultivation, or urban park.
2. Compact midrise 	Dense mix of midrise buildings (3–9 stories). Few or no trees. Land cover mostly paved. Stone, brick, tile, and concrete construction materials.	B. Scattered trees 	Lightly wooded landscape of deciduous and/or evergreen trees. Land cover mostly pervious (low plants). Zone function is natural forest, tree cultivation, or urban park.
3. Compact low-rise 	Dense mix of low-rise buildings (1–3 stories). Few or no trees. Land cover mostly paved. Stone, brick, tile, and concrete construction materials.	C. Bush, scrub 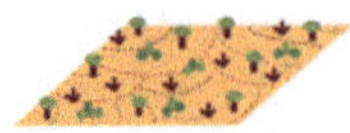	Open arrangement of bushes, shrubs, and short, woody trees. Land cover mostly pervious (bare soil or sand). Zone function is natural scrubland or agriculture.
4. Open high-rise 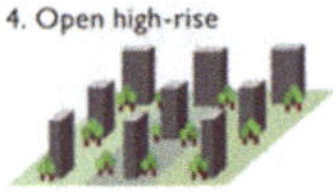	Open arrangement of tall buildings to tens of stories. Abundance of pervious land cover (low plants, scattered trees). Concrete, steel, stone, and glass construction materials.	D. Low plants 	Featureless landscape of grass or herbaceous plants/crops. Few or no trees. Zone function is natural grassland, agriculture, or urban park.
5. Open midrise 	Open arrangement of midrise buildings (3–9 stories). Abundance of pervious land cover (low plants, scattered trees). Concrete, steel, stone, and glass construction materials.	E. Bare rock or paved 	Featureless landscape of rock or paved cover. Few or no trees or plants. Zone function is natural desert (rock) or urban transportation.
6. Open low-rise 	Open arrangement of low-rise buildings (1–3 stories). Abundance of pervious land cover (low plants, scattered trees). Wood, brick, stone, tile, and concrete construction materials.	F. Bare soil or sand 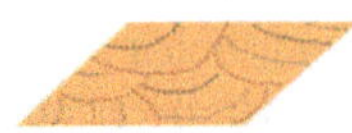	Featureless landscape of soil or sand cover. Few or no trees or plants. Zone function is natural desert or agriculture.
7. Lightweight low-rise 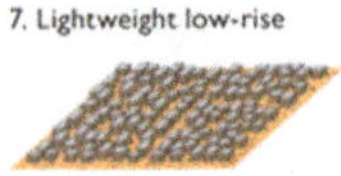	Dense mix of single-story buildings. Few or no trees. Land cover mostly hard-packed. Lightweight construction materials (e.g., wood, thatch, corrugated metal).	G. Water 	Large, open water bodies such as seas and lakes, or small bodies such as rivers, reservoirs, and lagoons.
8. Large low-rise 	Open arrangement of large low-rise buildings (1–3 stories). Few or no trees. Land cover mostly paved. Steel, concrete, metal, and stone construction materials.	**VARIABLE LAND COVER PROPERTIES** Variable or ephemeral land cover properties that change significantly with synoptic weather patterns, agricultural practices, and/or seasonal cycles.	
9. Sparsely built 	Sparse arrangement of small or medium-sized buildings in a natural setting. Abundance of pervious land cover (low plants, scattered trees).	*b. bare trees*	Leafless deciduous trees (e.g., winter). Increased sky view factor. Reduced albedo.
		s. snow cover	Snow cover >10 cm in depth. Low admittance. High albedo.
10. Heavy industry 	Low-rise and midrise industrial structures (towers, tanks, stacks). Few or no trees. Land cover mostly paved or hard-packed. Metal, steel, and concrete construction materials.	*d. dry ground*	Parched soil. Low admittance. Large Bowen ratio. Increased albedo.
		w. wet ground	Waterlogged soil. High admittance. Small Bowen ratio. Reduced albedo.

amplify their suitability by providing both high thermal loads and adequate installation surfaces. Conversely, historical compact districts with narrow street canyons, although thermally intense, revealed structural constraints related to limited pavement width, strong shading, and infrastructure complexity, which may reduce the technical feasibility of large-scale RPSC implementation. This highlights a critical

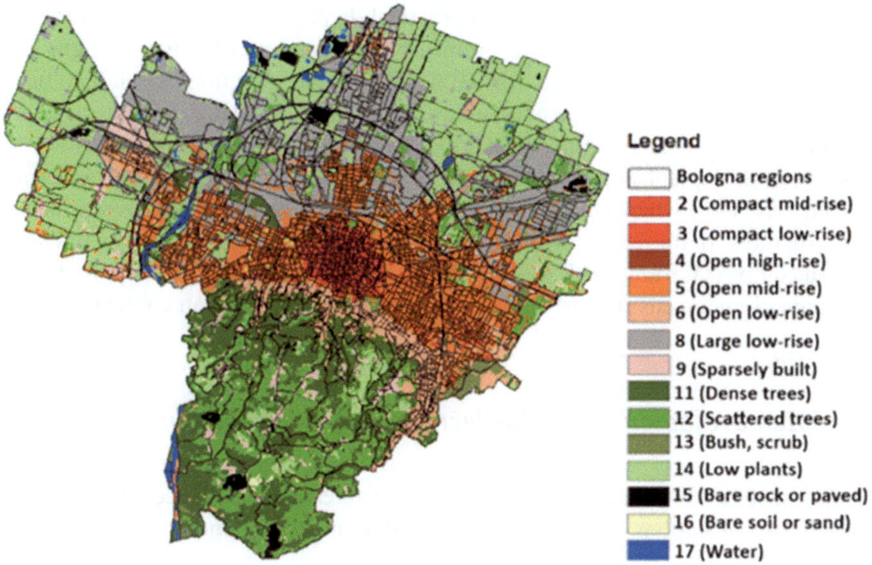

Fig. 4 Bologna LCZ map (generated by author)

Table 2 Division of LCZ Types into Heat Sources and Heat Sinks

LCZ Class	Description	Thermal Role	Effect on UHI
LCZ 2	Compact mid-rise	Source	Strong heat retention and release
LCZ 3	Compact low-rise	Source	Strong heat retention and release
LCZ 4	Open high-rise	Source	Heat absorption due to tall buildings
LCZ 5	Open mid-rise	Source	Moderate heat generation
LCZ 6	Open low-rise	Source	Moderate heat generation
LCZ 8	Large low-rise (industrial)	Source	Strong source due to paved/ roofed surfaces
LCZ 9	Sparsely built	Source	Weak source, limited vegetation
LCZ 15	Bare rock or paved	Source	Very strong heat source (impervious)
LCZ 16	Bare soil or sand	Source	Stores heat, releases at night
LCZ 11	Dense trees	Sink	Strong cooling via shade and evapotranspiration
LCZ 12	Scattered trees	Sink	Moderate cooling
LCZ 13	Bush / scrub	Sink	Moderate cooling, less than trees
LCZ 14	Low plants (grass, crops)	Sink	Cooling via evapotranspiration and soil permeability
LCZ 17	Water	Sink	Strong cooling, high heat capacity

trade-off between thermal priority and practical deployability that is likely to recur in other compact European cities. From a broader perspective, the Bologna case suggests that the optimal targets for RPSC systems are functionally defined urban corridors—such as commercial axes, industrial platforms, and high-traffic arterials—rather than urban density alone. This principle is transferable to other cities sharing similar morphological and climatic characteristics, particularly in Mediterranean and temperate regions, where compact built forms coexist with extensive paved transport infrastructures. In this sense, Bologna serves not merely as a single-city application, but as an empirical framework for identifying scalable RPSC priority typologies that can guide climate-sensitive urban design in a wide range of urban contexts.

3.2 Meteorological Data Collection and Preprocessing

At this stage, minimum and maximum temperature data are collected from urban meteorological stations and one or two rural stations as references [38].

$$T_{mean} = \frac{T_{min} + T_{max}}{2}, \; T_{range} = T_{max} - T_{min} \tag{1}$$

These derived formulas are commonly used in UHI studies to quantify diurnal thermal amplitudes and describe urban-rural temperature contrasts. For spatial consistency, the station coordinates are georeferenced and transformed into the Universal Transverse Mercator (UTM) coordinate system. As a result, these data can accurately overlap with the LCZ map and other remotely sensed data [37]. This produces a data set that provides a reliable basis for UHI modeling and increases the validity of subsequent prediction or interpolation models.

For each weather station, a circular buffer (e.g., radius 500 m) is created. The proportion of each local climate zone (LCZ) within the buffer is calculated as follows:

$$P_{LCZ_i} = \frac{Area\ of\ LCZ_i\ within\ buffer}{Total\ Buffer\ area} \tag{2}$$

This method follows established approaches in UHI-LCZ correlation analysis [37]. Each station is then assigned a source score that represents the combined thermal influence of the surrounding LCZs:

$$Source\_score = \sum_{i=1}^{10} w_i . P_{LCZ_i} \tag{3}$$

where w_i represents the weight associated with each LCZ as a heat source (positive) or heat sink (negative) [16, 25]. At this stage, the impact of urban morphology on local heat intensity is quantified and a basis is provided for selecting suitable locations for installing RPSC systems or creating additional green infrastructure [9].

3.3 Linear Regression Modeling of Temperature

In this step, the station temperatures are modeled as a function of the LCZ combination and the calculated source score:

$$y = \beta_0 + \sum_{i=1}^{n} \beta_i X_i + \varepsilon \tag{4}$$

where y is the observed temperature, X_i are LCZ fractions and the source score, β_i are regression coefficients, and ε represents random error. This model structure is commonly used in UHI prediction and urban thermal environment assessments [2].

The following metrics and Leave-One-Out Validation (LOO-CV) are used in this step to assess the model's performance. In order to provide an empirical foundation for UHI analysis and planning to lessen its effects, this stage aims to create a predicted relationship between urban morphology and observed temperature patterns [7].

$$RMSE = \sqrt{\frac{1}{N}\sum\nolimits_{i=1}^{N}\left(y_i - y_i'\right)^2}, MAE = \frac{1}{N}\sum\nolimits_{i=1}^{N}\left|y_i - y_i'\right|,$$

$$R^2 = 1 - \frac{\sum\left(y_i - y_i'\right)^2}{\sum(y_i - \bar{y}_i)^2} \tag{5}$$

3.4 Residual Spatial Interpolation and Kriging

At this stage, for each station *i*, the residual is calculated as follows:

$$Residual_i = T_{obs,i} - T_{pred,i} \tag{6}$$

conventional Kriging, where $_i$ are Kriging weights obtained from the spatial variogram, is used to spatially interpolate residuals to create a continuous temperature field.

$$Z(S_0) = \sum_{i=1}^{N} \lambda_i z(s_i) \tag{7}$$

In order to maximize source-sink balance and RPSC placement, this stage improves temperature surface predictions, locates localized UHI hotspots, and directs geographic interventions [4].

As previously mentioned, RPSC systems are strategically installed in areas identified as major heat sources to reduce surface temperatures and capture renewable energy for surrounding buildings. The solar energy absorbed by the pavement is converted into usable thermal or electrical energy:

$$Q_{collected} = \eta . I_{solar} . A_{pavement} \tag{8}$$

Where $Q_{collected}$ is collected heat energy, η is system efficiency, I_{solar} is incoming solar radiation, and $A_{pavement}$ is the collector surface area.

3.5 CFD Modeling for Microclimate and Ventilation

Navier–Stokes Equation (for incompressible flow):

$$\rho\left(\frac{\partial u}{\partial t} + (u.\nabla)u\right) = -\nabla p + \mu\nabla^2 u + F \tag{9}$$

Energy Equation:

$$\rho C_p\left(\frac{\partial T}{\partial t} + u.\nabla T\right) + k\nabla^2 T + S \tag{10}$$

CFD modeling supports design optimization for sustainable urban environments by evaluating the effects of green infrastructures and RPSC systems on ventilation, thermal comfort, and microclimate. This integrated method provides a comprehensive framework for understanding, predicting, and mitigating urban heat through an energy-efficient design strategy that is based on source-sink relationships. It combines RPSC technology with statistical modeling, CFD simulation, and LCZ-based spatial description.

4 Discussion and Implications

Integrating the source-sink theory with road pavement solar collector (RPSC) systems will provide a transformative path toward climate-resilient cities. This hybrid approach can be examined from two perspectives: mitigation and adaptation to urban climate change. In the mitigation perspective, implementing RPSC systems in areas that can be heat sources directly reduces surface temperatures and balances the thermal load between built-up areas and green cover. Because these systems extract thermal energy from pavements and convert it into thermal or electrical energy, they reduce the urban heat island (UHI) effect [33, 34].

On the other hand, the source-sink theory provides a conceptual framework for understanding the spatial heterogeneity of urban heat islands. It helps planners quantify and then map the spatial distribution of heat sources and sinks using the LCZ classification. When combined with geographic information systems (GIS), it is possible to identify optimal locations for RPSC facilities and then enhance green corridors and water bodies to maximize urban ventilation.

Another important consequence of this approach is increased energy sustainability. By directly harvesting waste heat or solar energy from pavements, RPSC systems can use renewable energy for heating, cooling, or ventilation, reducing dependence on fossil fuels. Furthermore, combining RPSC data with computational fluid dynamics (CFD) simulations allows for the assessment of airflow and temperature changes in the urban canopy layer, creating synergies between passive cooling (green infrastructure) and active energy harvesting (RPSC).

It is also worth noting that the integration of source-sink and RPSC frameworks is aligned with the Sustainable Development Goals (SDGs) 11 and 13, promoting resilient and climate-adapted infrastructure. Municipalities can use these models to prioritize investments in areas most exposed to heat stress and air pollution, especially during heat waves. From an economic perspective, the dual function of RPSC, cooling and energy generation, increases the cost-effectiveness of climate interventions. From a social perspective, urban cooling reduces health risks, alleviates energy poverty, and improves the liveability of dense urban areas. Overall, this interdisciplinary framework integrates physical climatology, renewable energy engineering, and spatial planning. It demonstrates how cities like Bologna can serve as testbeds for integrating microclimate modeling, LCZ-based thermal mapping, and innovative infrastructure, leading to scientific and scalable strategies for urban resilience. In contrast to many existing urban climate strategies that primarily focus on building energy efficiency, façade retrofitting, or the expansion of urban green spaces, the framework proposed here uniquely targets the extensive and largely underutilized network of road pavements, which constitute one of the dominant and persistent urban heat sources. While vegetation-based and building-centered interventions are essential, they often overlook the thermal and energetic potential of transport infrastructures, despite their continuous exposure to solar radiation and their large spatial coverage within dense cities. By explicitly integrating RPSC systems into the source–sink planning logic, this approach positions road surfaces not only as passive contributors to the UHI effect, but as active components of urban energy and climate regulation. Moreover, the thermal and electrical energy harvested by RPSC installations can be coupled with district heating and cooling networks, seasonal thermal storage, or smart energy management systems discussed elsewhere in this volume, creating multi-scale synergies between urban climate adaptation and urban energy transitions. In this sense, the proposed framework does not compete with conventional strategies, but complements them by adding a new infrastructural layer that transforms a neglected heat source into a strategic asset for climate-resilient and energy-efficient urban systems.

5 Concluding Remarks

This chapter investigates the application of source–sink theory to road pavement solar collector (RPSC) systems as a new framework for urban heat reduction and renewable energy generation. Using LCZ-based spatial analysis, statistical modeling, geostatistical interpolation, and CFD simulations, this study developed a methodology that connects urban morphology to thermal performance and technological adaptability. The discussion emphasized that source–sink theoryS provides a theoretical foundation for identifying spatial trade-offs between heat sources and sinks, whereas RPSC systems represent a practical instrument for transforming excess surface heat into useful energy.

These strategies enable cities to transition from reactive heat management toward proactive and energy-positive urban design. Although barriers to large-scale adoption remain—such as installation costs, maintenance requirements, and integration with existing infrastructure—the environmental and energetic potential of this technology justifies further research and pilot implementations. Future studies should therefore focus on integrating real-time monitoring data with high-resolution urban climate models in order to optimize RPSC design and quantify their long-term impacts on urban heat and energy balances.

More broadly, the integration of RPSC systems within a source–sink planning architecture points toward a new paradigm in urban climatology, in which urban infrastructure is no longer conceived as a passive recipient of climate stress, but as an active and regenerative component of the urban system. In this emerging vision, roads, pavements, and transport corridors are designed not only to support mobility, but simultaneously to mitigate climate risks and produce renewable resources. In this sense, cities such as Bologna can evolve into living laboratories of climate resilience, where surface cooling and energy harvesting become integral elements of future urban design, transforming cities from passive victims of heat into proactive and self-adaptive systems.

Acknowledgments I would like to express my deepest gratitude to my husband, whose love, patience, and unwavering support have been a constant source of strength throughout this journey. His encouragement in the most challenging moments and his belief in me, even when I doubted myself, made this possible. This book chapter is as much his as mine, and I dedicate it to him with love and gratitude.

References

1. Andoni, H., & Wonorahardjo, S. (2018). A review on mitigation technologies for controlling urban heat Island effect in housing and settlement areas. In IOP conference series: earth and environmental science. 152, 1, 012027. IOP Publishing
2. Oke TR (1982) The energetic basis of the urban heat Island. Q J R Meteorol Soc 108(455):1–24

3. Phelan PE, Kaloush K, Miner M, Golden J, Phelan B, Silva H III, Taylor RA (2015) Urban heat Island: mechanisms, implications, and possible remedies. Annu Rev Environ Resour 40(1):285–307
4. Mirzaei PA (2015) Recent challenges in modeling of urban heat Island. Sustain Cities Soc 19:200–206
5. Mohajerani A, Bakaric J, Jeffrey-Bailey T (2017) The urban heat Island effect, its causes, and mitigation, with reference to the thermal properties of asphalt concrete. J Environ Manag 197:522–538
6. Kong J, Zhao Y, Carmeliet J, Lei C (2021) Urban heat Island and its interaction with heatwaves: a review of studies on mesoscale. Sustainability 13(19):10923
7. Hou H, Wang R, Murayama Y (2019) Scenario-based modelling for urban sustainability focusing on changes in cropland under rapid urbanization: a case study of Hangzhou from 1990 to 2035. Sci Total Environ 661:422–431
8. Fu P, Weng Q (2016) A time series analysis of urbanization induced land use and land cover change and its impact on land surface temperature with Landsat imagery. Remote Sens Environ 175:205–214
9. Chen J, Zhao Z, Yang Y, Li C, Yin Y, Zhao X et al (2024) Metallogenic prediction based on fractal theory and machine learning in Duobaoshan area, Heilongjiang Province. *Ore Geology Reviews* 168:106030
10. Hokao K, Phonekeo V, Srivanit M (2012) Assessing the impact of urbanization on urban thermal environment: a case study of Bangkok metropolitan. International Journal of Applied 2(7):243–256
11. Labajo AL, Egido M, Martín Q, Labajo J, Labajo JL (2014) Definition and temporal evolution of the heat and cold waves over the Spanish central plateau from 1961 to 2010. Atmósfera 27(3):273–286
12. Santamouris M (2020) Recent progress on urban overheating and heat Island research. Integrated assessment of the energy, environmental, vulnerability and health impact. Synergies with the global climate change. Energ Buildings 207:109482
13. Oke TR, Johnson GT, Steyn DG, Watson ID (1991) Simulation of surface urban heat islands under 'ideal'conditions at night part 2: diagnosis of causation. Bound-Layer Meteorol 56(4):339–358
14. Feng J, Gao K, Khan H, Ulpiani G, Vasilakopoulou K, Young Yun G, Santamouris M (2023) Overheating of cities: magnitude, characteristics, impact, mitigation and adaptation, and future challenges. Annu Rev Environ Resour 48(1):651–679
15. Irfeey AMM, Chau HW, Sumaiya MMF, Wai CY, Muttil N, Jamei E (2023) Sustainable mitigation strategies for urban heat Island effects in urban areas. Sustainability 15(14):10767
16. Xiang Y, Cen Q, Peng C, Huang C, Wu C, Teng M, Zhou Z (2023) Surface urban heat Island mitigation network construction utilizing source-sink theory and local climate zones. Build Environ 243:110717
17. Bahi H, Mastouri H, Radoine H (2020) Review of methods for retrieving urban heat islands. Mater Today Proc 27:3004–3009
18. Bottyán Z, Bíróné Kircsi A, Szegedi S, Unger J (2005) The relationship between built-up areas and the spatial development of the mean maximum urban heat Island in Debrecen, Hungary. Int J Climatol 25:405–418
19. Bottyán Z, Unger J (2003) A multiple linear statistical model for estimating the mean maximum urban heat Island. Theor Appl Climatol 75(3):233–243
20. Lai LW, Cheng WL (2009) Air quality influenced by urban heat Island coupled with synoptic weather patterns. Sci Total Environ 407(8):2724–2733
21. Oke TR (1973) City size and the urban heat Island. Atmospheric Environment (1967) 7(8):769–779
22. Ren P, Meng Q, Zhang Y, Zhao L, Yuan X, Feng X (2015) An unmanned airship thermal infrared remote sensing system for low-altitude and high spatial resolution monitoring of urban thermal environments: integration and an experiment. Remote Sens 7(10):14259–14275

23. Sahashi K, Hieda T, Yamashita E (1996) Nitrogen-oxide layer over the urban heat Island in Okayama City. Atmos Environ 30(3):531–535
24. Tapper NJ, Tyson PD, Owens IF, Hastie WJ (1981) Modeling the winter urban heat Island over Christchurch, New Zealand. J Appl Meteorol Climatol 20(4):365–376
25. Zhao H, Zhang H, Miao C, Ye X, Min M (2018) Linking heat source–sink landscape patterns with analysis of urban heat islands: study on the fast-growing Zhengzhou City in Central China. Remote Sens 10(8):1268
26. Chen L, Fu B, Zhao W (2008) Source-sink landscape theory and its ecological significance. Front Biol China 3(2):131–136
27. Wu S, Mallick RB, Tao M, Chen B (2011) Investigation of heat harvesting and storage in asphalt pavements. Appl Therm Eng 31(17–18):2877–2887
28. Mallick RB, Chen B, Bhowmick S (2009) Harvesting heat energy from asphalt pavements: development of an experimental setup. J Transp Eng 135(6):527–536
29. Wang F et al (2021) Recent progress in energy harvesting from pavements: methods and applications. Renew Sust Energ Rev 135:110210
30. Gao Y et al (2010) Modeling of pavement solar collector system for the utilization of renewable energy. Energ Buildings 42(4):491–495
31. Lepore R, Van der Veken R, Verbruggen B, Helsen L (2018) Road pavement solar energy systems: performance analysis and future prospects. Renew Sust Energ Rev 92:733–747
32. Xie N, Zhang Y, Zhou Y (2022) Seasonal performance of pavement-integrated thermal energy systems: a review. Appl Energy 307:118166
33. He Y, Chen Z, Guo L (2020) A review of photovoltaic pavement applications: performance, durability, and challenges. Energy Convers Manag 209:112660
34. Wu M, Yang H, Li J (2021) Flexible photovoltaic pavements: materials, design, and field applications. Sol Energy Mater Sol Cells 230:111239
35. Zhao H, Deng Z, Xu X (2019) High thermal conductivity concretes for energy harvesting pavements. Constr Build Mater 225:743–755
36. Lowry WP (1977) Empirical estimation of urban effects on climate: a problem analysis. J Appl Meteorol Climatol 16(2):129–135
37. Stewart ID, Oke TR (2012) Local climate zones for urban temperature studies. Bull Am Meteorol Soc 93(12):1879–1900
38. Chen G, Chen Y, He H, Wang J, Zhao L, Cai Y (2023) Assessing the synergies between heat waves and urban heat islands of different local climate zones in Guangzhou, China. Build Environ 240:110434

Potential Strategies for Green Building Technology Adoption: Empirical Insights for Climate-Resilient Urban Development in Construction Sector

Bilal Hussain, Rui Feng Zhao, Syed Asif Ali Naqvi, and Atiq ur Rehman

1 Introduction

Construction sector has many economic and social benefits due to its backward and forward linkages with many allied industries. But it is also estimated to use 40% of total energy and is responsible for 30% of Greenhouse gas (GHG) emissions. This necessitates judicious use of resources in this sector to mitigate and adapt to climate change related hazards. Green Buildings (GBs) are structures for healthy and pleasant living and reducing global warming related risks [1]. GBs are constructions and practices that are resource efficient and accountable throughout building's life cycle from design to operation and Green Building Technologies (GBTs) are specific methods and material [2].

Green building technologies (GBTs) are being advocated as a promising solution to reduce the negative environmental externalities and have become increasingly crucial in fostering climate-resilient cities [3]. GBTs are designed to use energy efficiently, conserve rainwater and sustainable material usage in building structures. These include use of solar energy systems and other renewable sources for resource conservation, high-performance insulation and recyclable material in building design and smart technologies and integration of IoT for energy efficiency [4]. Accordingly Green Buildings reconcile benefits accrued from economic, social and environmental dimensions [5]. GBTs also have a positive impact on human well-being and comfort [6].

B. Hussain · R. F. Zhao (✉)
Faculty of Management and Economics, Kunming University of Science and Technology, Kunming 650000, China
e-mail: 20110227@kust.edu.cn

B. Hussain · S. A. A. Naqvi · A. Rehman
Department of Economics, Government College University Faisalabad, Faisalabad 38000, Pakistan

A. Cheshmehzangi et al. (eds.), *Design Innovation for Climate-Resilient Cities and Built Environments*, Urban Sustainability, https://doi.org/10.1007/978-981-92-2297-1_10

Despite having great potential for climate mitigation and adaptation, GBTs remain under-utilized in developing economies [7]. There has been a lot of research on the potential barriers of effective uptake of GBTs notably lack of financial incentives, ineffective regulatory framework and high upfront costs (Chan et al. 2018). Barriers are grouped in government related, market related, human related, knowledge related, and cost and risk related. There are some country/region specific barriers particularly bifurcating between developed and developing economies.

Pakistan is among top five most affected countries by climate change [8]. Urbanization is increasing demand for housing and construction, but it comes with associated negative environmental externalities of old construction methods. Pakistan's energy consumption will increase many folds in coming years and will contribute to GHG emissions and other pollutants causing adverse impacts on climate and public health [9]. Nonetheless, Pakistan is committed to Paris Agreement and SDG11 for sustainable urban development and resilient cities. Therefore, Pakistan' National Climate Change Policy carries specific national targets for diffusion of GBTs as part of its low carbon footprint focusing on adaption and mitigation. The Pakistan Engineering Council released Green Building Code in 2023 to provide regulatory frameworks for eco-friendly practices in the construction sector. It entails guidelines for building designs, construction and operational stages.

However, uptake of GBTs in Pakistan is at nascent stage [10] due to contextual factors. Stakeholders in the construction sector, including architectures, builders, government departments and general masses, use mostly informal practices. They lack proper knowledge of latest and trending technologies, regulations and incentives available to them. Barriers mentioned in the literature also hinder progress toward GBTs adoption especially high upfront costs, lack of government incentives and lack of financing [11]. These barriers coupled with investment risks need some triggers through institutional support for speedy diffusion of GBTs. Stakeholders' involvement is necessary for any policy or strategy to increase uptake of GBTs. This study explores how and which strategies work best and for whom in a local context. By quantifying different policy mixes and sorting out most effective strategy will help minimize policy risk and efficient use of economic resources.

2 Data and Methodology

The research is conducted in construction industry of Pakistan. The appropriate selection of data collecting approach in information sciences study is one of the most crucial elements in moving research ahead. Using the proper data collecting approach assured the validity of the data of the research conducted in this area of the study. The data collecting methodology is a critical step in quantitative analysis. The primary data collection method for this research was questionnaires, and it was utilized to get solicitation from experts. In the GB literature, this survey method has been a popular approach for analyzing the factors impacting the adoption of GBTs. It was used to collect stakeholders' opinions in CI to discuss the numerous barriers, drivers, strategies and their behavior toward adopting GBTs in Pakistan. This research uses closed-ended questions built on a five-point Likert scale because it provides transparent, simple-to-understand findings and widely utilized in GBTs research. Furthermore, the items ranged from 1 (strongly disagree) to 5 (strongly agree). Respondents were enquired to rate the questions based on the scale.

The questionnaire is divided into ten sections. Section 1 comprises questions on the respondent's personal information. Age, education, type of CI, role in the CI, knowledge about GBTs, etc., are all included in this section. Furthermore, Sect. 2 contains GB technologies being used in CI in Pakistan. Section 3 consists of questions regarding stakeholders' adoption behavior toward GBTs in Pakistan. Likewise, Sects. 4, 5, and 6 contain the questions regarding critical barriers, main drivers, and potential approaches toward adopting GBTs in Pakistan, respectively. Furthermore, the questions regarding the stakeholders' adoption behavior toward GBTs in the context of technology, organizational, environmental, and economics are presented in Sects. 7, 8, 9, and 10, respectively.

The study was carried out by personally delivering questionnaires to all target stakeholders (e.g., architects, engineers, suppliers, consultants, contractors, and developers etc.) in major cities of Punjab province, including Lahore, Faisalabad, Rawalpindi, Gujranwala, and Multan. These cities were chosen based on the fact that the CI has significant markets there. The estimated sample size was 384 questionnaires delivered to GBT stakeholders. However, 349 valid responses were attained with a response rate of 90.88%. A similar high response rate was also found in previous studies such as [5] (85.8%), [12] (90%), [13] (80%), and [14] (92%), supporting the study's response rate. The fact that the researcher specifically targeted stakeholders interested in the study may be responsible for the relatively high response rate.

Furthermore, a list of 5 promotional strategies and their 15 items is presented in Table 1. In this respect, awards and acknowledgment have two items, whereas education and R&D support and incentives toward GBTs have three and four items, respectively. Moreover, government standards, regulations, publicity programs, and awareness have 3, 3 items, respectively. All the promotional strategies are identified from various previous studies (such as., [6], 11 15).

2.1 *Partial Least Square Structural Equation Modeling (PLS-SEM)*

This research used smart Partial Least Squares (smart PLS) method to test all the study hypotheses. Smart PLS is a statistical technique that can analyze relationships between latent variables and their indicators [16]. It is a variant of the PLS regression technique, which is an alternative to ordinary least squares (OLS) regression when dealing with data that has collinearity issues or when the number of predictors exceeds

Table 1 List of promotional strategies with their items

Variable	Items	Definition
Awards and acknowledgment	AAS1	Efficient GBTs adopters should be acknowledged and rewarded
	AAS2	Government should categorize certificates for different levels of GBTs adoption
Education and R&D support	ERDS1	For effective implementation of GBTs, institutional and R&D frameworks should be promoted
	ERDS2	Certification schemes for the adoption of GBTs should be introduced
	ERDS3	Green procurement should be promoted
Incentives toward GBTs	GBTIS1	Government should provide insurance to adopters of GBTs
	GBTIS2	Market-based and financial incentives should be provided to GBTs adopters (such as tax holidays, relaxed duties on the import of GBTs
	GBTIS3	Government and financial institutions should provide Low-interest loans, subsidies, and easy credit policies for GBTs adopters
	GBTIS4	Government should promote public–private partnerships
Government standards and regulations	GSRS1	Local government authorities and teams should be competent, proactive, and active in successfully promoting GBTs
	GSRS2	Government should push relevant ministries to establish standards and regulations
	GSRS3	Mandatory Green Building codes and regulations should be available
Publicity programs and Awareness	PPAS1	Executive management should develop benchmarks for the successful promotion of GBTs
	PPAS2	Environmental awareness should be promoted through seminars, workshops, and conferences
	PPAS3	GBTs publicity should be through social media, print media, television programs, etc. programs)

the number of samples (Hair et al. 2006). Thus, all the study' variables are analyzed using PLS method.

Smart PLS offers several advantages over traditional PLS, such as improved estimation accuracy and robustness in the presence of outliers, better interpretation of results through the use of standardized coefficients, the ability to control for measurement errors, and the ability to perform multiple analyses in one run [17]. Smart PLS can be used to investigate the relationships between two or more latent variables and their indicators. The technique uses an equation-based approach to estimate the effects of latent variables on their indicators. Data analysis in the smart PLS method consists of two models: the measurement model and the structural model. The measurement Model demonstrates how latent variables, which cannot be measured directly and must be expressed by other variables, are represented by the observed variables. However, the structural model shows the estimated power between these latent variables [18]. Therefore, this model may be considered to explain the connection between latent variables. The PLS approach is designed as a recursive computation model and each dependent variable in the link between latent variables is called a causal chain system. However, the collection of function equations will result in the formation of a latent variable.

3 Results and Discussion

This section fetches the main findings and relevant discussion. Data analysis is divided into two parts, measurement model and structural model. First analysis estimates latent variables which cannot be measured directly and second one explains the estimated power between these latent variables.

The PLS-SEM measurement model is shown in Fig. 1. This figure and Table 2 show the factor loadings. Previous research suggests that factor loadings (FL) should not be less than 0.7 (Hair et al. 2017). The measurement model findings reveal that awards and acknowledgment have FL from 0.935 to 0.953, while education and R&D support have FL ranging from 0.766 to 0.952. Nevertheless, incentives toward GBTs have FL from 0.794 to 0.837; government standards and regulations have FL ranging from 0.773 to 0.829; and publicity programs and awareness have FL from 0.799 to 0.890. Finally, SAB has factor loadings ranging from 0.777 to 0.901. As a result, all scale items have factor loadings greater than 0.7.

Moreover, the validity and reliability of the coefficients of latent constructs were utilized to evaluate the measurement model's quality. It is plausible for structural analysis based on the findings of the latent constructs (Hair et al. 2017). Specifically, the CR values ranged from 0.839 to 0.942, higher than the specified top limit of 0.70. [16]. Moreover, the CA values fluctuated from 0.821 to 0.878, higher than the recommended limit of 0.7 [19]. Moreover, the AVE coefficients for all variables were higher than 0.50, ranging from 0.635 to 0.891, demonstrating the model's latent variables' convergent validity and reliability (Hair et al. 2021).

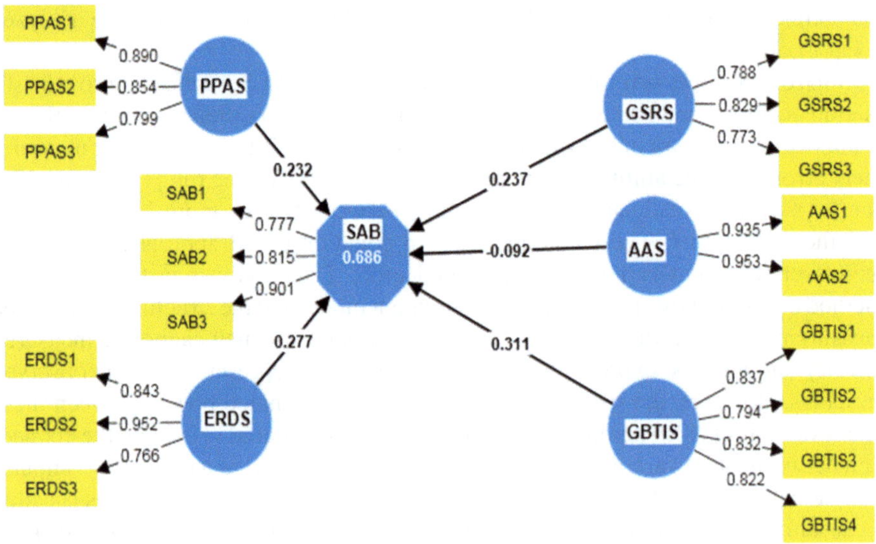

Fig. 1 Path diagram of the measurement model

Table 2 Items with VIF, outer loadings, CA, rho_a, CR, and AVE

Construct	No. of items	VIF	Outer loadings	CA	rho_a	CR	AVE
Awards and acknowledgment	AAS1	2.578	0.935	0.878	0.894	0.942	0.891
	AAS2	2.578	0.953				
Education and R&D support	ERDS1	3.271	0.843	0.815	0.849	0.892	0.735
	ERDS2	4.416	0.952				
	ERDS3	1.682	0.766				
Incentives toward GBTs	GBTIS1	4.677	0.837	0.839	0.840	0.892	0.675
	GBTIS2	3.772	0.794				
	GBTIS3	4.660	0.832				
	GBTIS4	3.948	0.822				
Government standards and regulations	GSRS1	1.974	0.788	0.721	0.727	0.839	0.635
	GSRS2	2.144	0.829				
	GSRS3	1.187	0.773				
Publicity programs and awareness	PPAS1	2.110	0.890	0.805	0.811	0.885	0.720
	PPAS2	1.950	0.854				
	PPAS3	1.501	0.799				
Adoption of GBTs	SAB1	1.644	0.777	0.777	0.781	0.871	0.693
	SAB2	1.634	0.815				
	SAB3	2.326	0.901				

Fornell-Larcker, HTMT criterion, and cross-loadings were applied to gauge discriminant validity [20, 21]. The square root of all constructs' AVEs in the matrix diagonal is greater than the relevant correlations in the corresponding rows and columns, as shown in Table 3, demonstrating the reflecting model's quality. For example, the AVE for incentives toward GBTs (0.821) is larger than the equivalent row (0.747 and 0.618) and column correlations (0.592, 0.533, and 0.726). As a result, the latent variables in the research model vary, indicating the quality of the measured construct. Moreover, all constructs' HTMT values are smaller than the suggested limit of HTMT 0.90 [21], indicating that the latent variables employed in the research were conceptually different.

The cross-loadings (Table 4) reveal that the variable has a greater connection than other constructs. These data support discriminant validity and demonstrate a low correlation between variables.

After evaluating the measurement model, the research employs the structural model (Hair et al. 2021). The structural model investigates the connections between the variables under consideration. For relationships between variables to be significant, the t-statistic values must be more than 1.96, and the p-values must be less than 0.05. Using a structural model, Fig. 2 depicts the link between independent and dependent variables. The structural model is evaluated using the following criteria: collinearity (VIF), path coefficient significance, and R^2 values [7]. A bootstrapping approach with 5,000 re-samples is used to assess the provided hypotheses. It is a non-parametric re-sampling procedure that is performed after randomly replacing the original sample to provide many samples for the bootstrap test (Hair et al. 2011).

Table 3 Findings of discriminant validity

Fornell-Larcker criterion						
	AAS	ERDS	GBTIS	GSRS	PPAS	SAB
AAS	***0.944***					
ERDS	0.682	***0.857***				
GBTIS	0.618	0.747	***0.821***			
GSRS	0.530	0.549	0.592	***0.797***		
PPAS	0.425	0.584	0.533	0.586	***0.848***	
SAB	0.514	0.713	0.726	0.661	0.660	0.833
HTMT ratio						
	AAS	ERDS	GBTIS	GSRS	PPAS	SAB
AAS						
ERDS	0.799					
GBTIS	0.716	0.893				
GSRS	0.681	0.678	0.739			
PPAS	0.503	0.711	0.643	0.729		
SAB	0.613	0.882	0.895	0.843	0.835	

Table 4 Loadings and cross-loadings

	AAS	ERDS	GBTIS	GSRS	PPAS	SAB
AAS1	***0.935***	0.647	0.548	0.518	0.383	0.444
AAS2	***0.953***	0.641	0.614	0.486	0.417	0.520
ERDS1	0.579	***0.843***	0.529	0.508	0.486	0.515
ERDS2	0.708	***0.952***	0.720	0.641	0.606	0.719
ERDS3	0.445	***0.766***	0.653	0.231	0.392	0.573
GBTIS1	0.546	0.709	***0.837***	0.507	0.528	0.612
GBTIS2	0.452	0.505	***0.794***	0.436	0.312	0.572
GBTIS3	0.513	0.696	***0.832***	0.503	0.527	0.613
GBTIS4	0.518	0.536	***0.822***	0.497	0.373	0.585
GSRS1	0.434	0.355	0.457	***0.788***	0.263	0.475
GSRS2	0.522	0.400	0.409	***0.829***	0.440	0.420
GSRS3	0.343	0.519	0.520	***0.773***	0.634	0.631
PPAS1	0.381	0.582	0.485	0.521	***0.890***	0.606
PPAS2	0.375	0.505	0.413	0.551	***0.854***	0.533
PPAS3	0.325	0.391	0.456	0.420	***0.799***	0.535
SAB1	0.361	0.537	0.562	0.490	0.539	***0.777***
SAB2	0.513	0.661	0.685	0.555	0.502	***0.815***
SAB3	0.397	0.572	0.556	0.600	0.608	***0.901***

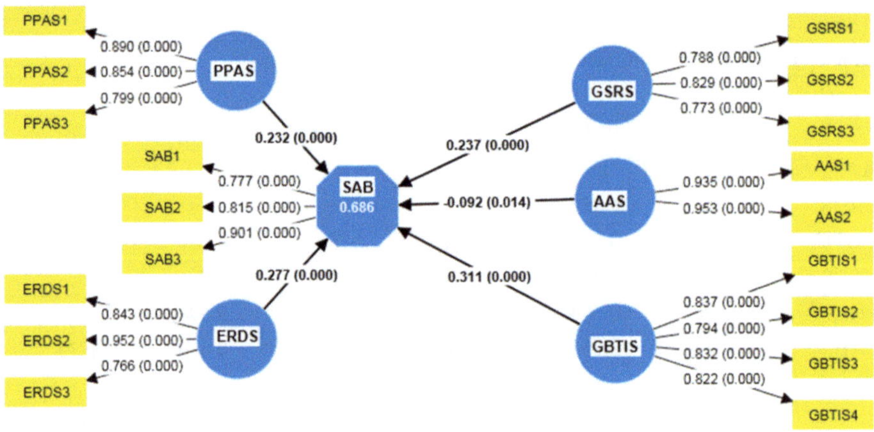

Fig. 2 Path diagram for structural model

To begin, all predictor constructs' VIF values were examined to look for collinearity issues [3] established the most stringent criterion: VIF values should be less than 5. All construct values are lower than the reference value, as shown in Table 2. As a result, the study's model has no collinearity issues. Second, the research examined the path coefficients. These path coefficients show the strength of the correlations between the explanatory factors and the variables to be explained. Table 5 displays the path coefficients' findings. In this regard, it is demonstrated that five of the five postulated links in the theoretical model are relevant.

Figure 2 also reports the findings of the structural model. According to the results, awards and acknowledgment ($\beta = -0.092$; $p < 0.05$), education and R&D support ($\beta = 0.277$; $p < 0.01$), incentives toward GBTs ($\beta = 0.311$; $p < 0.01$), government standards and regulations ($\beta = 0.237$; $p < 0.01$), and publicity programs and awareness ($\beta = 0.232$; $p < 0.01$) were the significant strategies of adoption of GBTs. According to the research findings, government standards and regulations are the most important method for promoting GBT adoption in Pakistan, followed by education. This may explain why socio-cultural and government barriers were deemed the second most noteworthy barrier to GBT adoption because it is reasonable to assume that government standards and regulations, such as the lack of green building policies may greatly assist in overcoming socio-cultural and government barriers.

This research measured the impact of government standards and regulations on boosting the use of GBTs and discovered that government standards and regulations had a considerable positive impact (path coefficient of 0.237). Government standards and regulations would significantly impact GBT adoption because they would put enterprises and stakeholders under regulatory pressure to use GBTs. According to [22], regulatory pressure is the primary motivation for stakeholders to adopt GBTs and practices. According to [23], one of the activities within the building sector is that "if you don't regulate, people won't start doing it" is the adoption of GBTs and practices. The current research suggests that by enacting obligatory green construction laws and regulations, the Pakistani government might considerably increase the

Table 5 Result of the structural model

Hypotheses		Path coefficient	STD	T-stat	CI Levels		P value	f^2	Support
					2.5%	97.5%			
H3a	AAS -> SAB	-0.092	0.03	2.45	-0.168	-0.020	0.014	0.013	Yes
H3b	ERDS -> SAB	0.277	0.04	6.17	0.189	0.366	0.000	0.081	Yes
H3c	GBTIS -> SAB	0.311	0.03	8.23	0.238	0.387	0.000	0.118	Yes
H3d	GSRS -> SAB	0.237	0.03	7.50	0.174	0.297	0.000	0.093	Yes
H3e	PPAS -> SAB	0.232	0.03	7.37	0.173	0.298	0.000	0.095	Yes

usage of GBTs. When faced with required government mandates, stakeholders would have little option but to implement GBTs in their projects to avoid fines and penalties for noncompliance. Additionally, due to a lack of enforcement, construction stakeholders in several countries have reported poor understanding and use of many of the green laws and regulations established and approved by policymakers (Mulligan et al. 2014). As a result, once adopted, improved enforcement of green building rules is critical to increasing GBTs adoption.

Similarly, green building rating systems are significant techniques for promoting GBT usage. Green building rating systems must be designed and implemented to encourage the use of GBTs. On the other hand, Pakistan does not yet have its own green building rating systems and hence relies on the LEED grading systems developed in the United States. While these worldwide rating systems may be valuable in boosting GBT adoption in Pakistan, creating localized grading systems that consider local sustainability concerns may be more beneficial. The government and other key stakeholders should assist the Pakistan Green Building Council to develop suitable green building grading systems. This might potentially be a fruitful subject for researchers to investigate. Another important marketing approach is the availability of knowledgeable and proactive GBTs promotion teams and local governments. According to [24], strong champions are also important in encouraging innovations, particularly green innovation, in the construction business.

Moreover, this study found that publicity programs, awareness, and incentives toward GBTs are significantly positively linked to GBTs adoption. The result also indicates that publicity and awareness programs are effective strategies for adopting GBTs in Pakistan. The significance of this strategy was also reinforced by references [11], where increased media exposure was a key promotion strategy for GBTs and green building standards adoption. Publicity, also known as public relations, is a marketing approach that may assist in establishing a favorable image for a product, encourage people to use the product, communicate the product's advantages, raise awareness, and generate demand for the product [25].

the importance of providing GBT-related education and R&D support to stakeholders in encouraging GBTs adoption cannot be overstated. Darko et al. [11] recognized instructional programs for stakeholders as one of the top five most significant tactics for catalyzing the adoption of green building rules in Ghana, a developing nation. Since stakeholders are major players in the adoption and promotion processes, educating and training them about GBTs is critical to influencing and pushing GBTs adoption in the CI. Developers, for example, play substantial and essential roles in the adoption of GBTs and practices. Developers are not only significant decision-makers in the adoption of GBTs, but their use of GBTs also influences R&D, contractors' construction strategy, and manufacturer investments. Similarly, [26] believe developers are major decision-makers in implementing green practices within the business since they are investors.

Given these considerations, creating and executing effective GBT-related contractors and programs to increase developers' knowledge, awareness, and experience in GBTs will undoubtedly significantly influence GBTs adoption. Similarly, because developers have a great deal of influence over firms and individual practitioners in the construction industry in a way that promotes innovation [24]. Adoption of GBTs would progressively become an industry-wide acknowledged practice in this manner. This might explain why GBT-related education and R&D support for stakeholders have the second-largest influence on GBTs adoption.

In contrast, awards and acknowledgments strategies are not positively linked to the stakeholder' adoption behavior toward the GBTs in Pakistan. This result suggests that this promotional strategy group would have little impact on GBTs adoption in the Pakistani CI. The results might be related to effective green building laws and incentives being seen as more significant than measures relating to awards and acknowledgments in promoting the use of GBTs.

4 Conclusions

Pakistan needs enormous housing and construction projects for its 225 million populations. It faces dual challenge of rising housing demand and resource stress old construction methods. Therefore, GBTs adoption and speedy diffusion is not an option for Pakistan now. It must encourage, facilitate and regulate this uptake for mitigation and adaptation of climate change risks. This study highlighted five major strategies from stakeholders' perspective. Government standards and regulations are deemed pivotal and most important policy lever to standardize and normalize this uptake. Financial incentives and subsidies are needed to minimize investment risks and make projects economically viable. Education and R&D promote innovative technologies and align them to domestic needs. Certification programs with stakeholders' consultation should be devised. Publicity programs through workshops and seminars, and awareness campaigns using latest communications means to socially normalize this practice of GBTs adoption are needed at all levels of stakeholder engagement. Individual recognition and awards, although not significant in our sample, will have social network effect in the long run. Policy makers and town planners now realize and incorporate all international standards and regulations in future policy frameworks. Coordination among different stakeholders and even in different government departments like building departments, climate change, environment protection, and urban planning units is needed to make cities livable and climate resilient.

References

1. He BJ (2019) Towards the next generation of green building for urban heat island mitigation: zero UHI impact building. Sustain Cities Soc 50:101647
2. Olubunmi OA, Xia PB, Skitmore M (2016) Green building incentives: a review. Renew Sustain Energy Rev 59:1611–1621

3. Ali F, Rasoolimanesh SM, Sarstedt M, Ringle CM, Ryu K (2018) An assessment of the use of partial least squares structural equation modeling (PLS-SEM) in hospitality research. Int J Contemp Hosp Manag 30(1):514–538
4. Ahmad T, Thaheem MJ, Anwar A (2016) Developing a green-building design approach by selective use of systems and techniques. Archit Eng Des Manag 12(1):29–50
5. Azeem S, Naeem MA, Waheed A, Thaheem MJ (2017) Examining barriers and measures to promote the adoption of green building practices in Pakistan. Smart Sust Built Environ 6(3):86–100
6. Wang T, Seo S, Liao PC, Fang D (2016) GHG emission reduction performance of state-of-the-art green buildings: review of two case studies. Renew Sustain Energy Rev 56:484–493
7. Reyes-Menendez A, Palos-Sanchez PR, Saura JR, Martin-Velicia F (2018) Understanding the influence of wireless communications and Wi-Fi access on customer loyalty: a behavioral model system. Wirel Commun Mob Comput 2018:1–16
8. Germanwatch (2025) Climate risk index 2025. https://www.germanwatch.org/en/cri Last accessed: September, 2025
9. Khan MA, Wang CC, Lee CL (2021) A framework for developing green building rating tools based on Pakistan's local context. Buildings 11(5):202. https://doi.org/10.3390/buildings11050202
10. Portnov BA, Trop T, Svechkina A, Ofek S, Akron S, Ghermandi A (2018) Factors affecting homebuyers' willingness to pay green building price premium: evidence from a nationwide survey in Israel. Build Environ 137:280–291
11. Chan AP, Darko A, Ameyaw EE, Owusu-Manu DG (2017) Barriers affecting the adoption of green building technologies. J Manag Eng 33(3):04016057
12. Hopkins EA (2016) Barriers to adoption of campus green building policies. Smart Sust Built Environ 5(4):340–351
13. Onubi HO, Yusof NA, Hassan AS, Bahdad AAS (2021) Analyzing the mediating effect of economic performance on the relationship between green construction practices and health and safety performance in Nigeria. Environ Sci Pollut Res 28(27):36598–36610
14. Mao T, Liu Y, Chen WQ, Li N, Dong N, Shi Y (2022) Quantifying spatiotemporal dynamics of urban building and material metabolism by combining a random forest model and GIS-based material flow analysis. Front Earth Sci 10:944865
15. Fu Y, Dong N, Ge Q, Xiong F, Gong C (2020) Driving-paths of green buildings industry (GBI) from stakeholders' green behavior based on the network analysis. J Clean Prod 273:122883
16. Bagozzi RP, Yi Y (1988) On the evaluation of structural equation models. J Acad Mark Sci 16:74–94
17. Tubbesing S (2017) Smart partial least squares: equations and case studies. Int J Bus Res Manag 8(4):14–30
18. Risal S, Khariry M, Asmawatiy C, Saputra RW, Sulandjari S, Elizabeth R (2023) Building trust through quality of service mediated with satisfaction to acceptors of family planning in the province of South Kalimantan. Heliyon, e13142
19. Nunnally JC (1975) Psychometric theory—25 years ago and now. Educ Res 4(10):7–21
20. Fornell C, Larcker DF (1981) Evaluating structural equation models with unobservable variables and measurement error. J Mark Res 18(1):39–50
21. Henseler J, Ringle CM, Sarstedt M (2015) A new criterion for assessing discriminant validity in variance-based structural equation modeling. J Acad Mark Sci 43:115–135
22. Shen L, Yan H, Fan H, Wu Y, Zhang Y (2017) An integrated system of text mining technique and case-based reasoning (TM-CBR) for supporting green building design. Build Environ 124:388–401
23. Gou Z, Lau SSY, Prasad D (2013) Market readiness and policy implications for green buildings: case study from Hong Kong. J Green Build 8(2):162–173
24. Blayse AM, Manley K (2004) Key influences on construction innovation. Constr Innov 4(3):143–154
25. Belch G,Belch M (2019) Advertising and promotion–an integrated marketing communication perspective (6-month. McGraw-Hill Education

26. Hu X, Xia B, Skitmore M, Buys L, Hu Y (2017) What is a sustainable retirement village? Perceptions of Australian developers. J Clean Prod 164:179–186
27. Darko A, Chan AP (2017) Review of barriers to green building adoption. Sustain Dev 25(3):167–179
28. Darko A, Chan APC (2018) Strategies to promote GBTs adoption in developing countries: The case of Ghana. Build Environ 130:74–84
29. Hair Jr, JF, Hult GTM, Ringle CM, Sarstedt M (2021) A primer on partial least squares structural equation modeling (PLS-SEM). Sage publications
30. Hair JF (2014) A primer on partial least squares structural equation modeling (PLS-SEM). Sage
31. Hussain B, Naqvi SAA, Anwar S (2023) Analyzing the impact of critical barriers on the stakeholder's adoption behavior toward green building technologies in Pakistan. Intern J Manag Res Emerg Sci 13(1)
32. Hussain B, ur Rehman A, Naqvi SAA (2025) Role of green building technologies in achieving climate-resilient cities. In *Designing healthy buildings and communities: shaping a climate-resilient future* (pp 85–96). Singapore: Springer Nature Singapore
33. Thackeray R, Neiger BL, Hanson CL (2007) Developing a promotional strategy: important questions for social marketing. Health Promot Pract 8(4):332–336

Furthering a Systemic Approach to Resilience: Planning, Infrastructure, and Applications of Adaptation

Ling Almoubayyed, Alec Barnes, Megan Campbell, Jenny Hamel, Bethanie Martin, Alicia Adams, and Katrina Kelly-Pitou

1 Introduction

Traditional approaches to resilience have focused on the ability to "bounce back" from individual disruptions that include floods, storms, or economic downturns treated as isolated events with localized solutions. However, the increasing complexity and interconnection of modern urban systems demands a fundamental shift in how we conceptualize and implement resilience strategies. Climate change, aging infrastructure, economic volatility, and social inequities do not operate in isolation; they form networks of risk that can rapidly overwhelm communities unprepared for systemic failure.

The economic losses from disasters have been trending upward for decades, which reflect not only intensifying climate extremes but also our growing vulnerability as development expands, populations concentrate, and material wealth increases. This trajectory underscores a critical gap between different levels of governmental organizations. While individual sectors, for example the transportation, energy, and housing departments, have developed their own risk management strategies, few frameworks exist to address the interdependence between these systems or to build adaptive capacity at the scale where most planning decisions are made at the municipal level. Whether wildfires in California, ice storms in Michigan, or hurricanes in Louisiana, when disaster strikes, it impacts these systems simultaneously, exposing the weak points that exist when levels of government are not working together for holistic resilience.

This chapter presents a system-based approach to resilience that moves beyond reactive recovery toward proactive transformation. Drawing from SmithGroup's

L. Almoubayyed · A. Barnes · M. Campbell · J. Hamel · B. Martin · A. Adams ·
K. Kelly-Pitou (✉)
IMPACT (Integrated Multidisciplinary Performance Analytics and Climate-impact Team), SmithGroup, USA
e-mail: Katrina.Kellypitou@Smithgroup.com

A. Cheshmehzangi et al. (eds.), *Design Innovation for Climate-Resilient Cities and Built Environments*, Urban Sustainability, https://doi.org/10.1007/978-981-92-2297-1_11

work with the RRC Resilience Toolkit and project applications in Michigan communities, we demonstrate how resilience can be operationalized through the four interconnected pillars of people, place, economy, and infrastructure. Rather than treating these as separate sectors, this framework recognizes that community resilience emerges from the dynamic relationships between social networks, built environments, economic systems, and physical infrastructure.

The urgency of this approach has been validated through direct experience working with different regional and local organizations. Communities like Midland and Alpena, Michigan, which participated in resilience planning processes, subsequently faced major disasters including catastrophic failure of two dams and weeks-long power outages due to ice storms lasting several days. These events revealed how shocks cascade across all four pillars of the resilience framework simultaneously, disrupting not just physical systems but also economic activity, social cohesion, and governance capacity. Such experiences highlight why resilience cannot be achieved through isolated interventions but requires integrated, regionally coordinated strategies rooted in systems thinking.

This chapter also explores how resilience theory translates into innovative infrastructure solutions, particularly through emerging opportunities in high-performance computing (HPC) and district energy systems. As communities navigate through both climate adaptation and economic development pressures, such integrated approaches offer pathways to align infrastructure investments with long-term resilience goals while generating immediate public value.

2 Overview

2.1 Urban Resilience as a Systems Condition

The economic losses attributed to disaster events and natural catastrophes have trended upward since the 1980s (NOAA n.d.); in addition to intensifying weather extremes driven by climate change, this rise relates to increasing standards of living and material wealth, greater investment in development, expansion across the wildland-urban interface, and the densification of population centers. This can be seen as an increasingly interconnected web of connectivity among our natural, social, and economic systems. To combat this trend, cities must begin planning for resilience, which involves understanding the risks associated with growth and development and creating actionable steps toward adaptation to the changing climate and social circumstances to which the community is exposed.

Urban resilience is best understood as a systems condition that emerges from interactions among social, ecological, and technical domains, not a single sector's performance. Drawing on socio-ecological systems (Holling, Folke), planning and disaster risk-reduction scholarship (Davoudi; Cutter; UNISDR), engineered infrastructure and network science (Rindaldi, Buldyrev; O'Rourke) and energy transition research (Kelly; Kelly-Pitou), resilience can be framed as a dynamic capacity where anticipation, absorption, adaptation, and transformation need to be equally distributed across people, place, economy, and infrastructure.

2.2 *The Governance and Planning Gap*

Despite advances in resilience theory, practice has lagged. Many governments still pursue sector-specific risk management strategies—transportation departments address flooding, utilities manage grid vulnerabilities, housing agencies address affordability, without coordinating across interdependencies. Scholars argue that fragmented governance reproduces vulnerability, particularly at the municipal scale where planning decisions most directly shape community outcomes.

Urban resilience research further highlights a governance gap: while disasters often manifest locally, the resources, policies, and coordination needed to address them span regional, state, and federal scales. This misalignment creates a pressing need for frameworks that enable municipalities to both anticipate cascading risks and collaborate across boundaries. This was seen notably in the aftermath of Hurricane Katrina, particularly in areas like the Lower Ninth Ward, a low-income community in New Orleans where recovery resources have never been abundant. After two decades, despite commendable efforts by residents to rebuild their community, the billions in federal dollars spent on statewide recovery and outside philanthropic investment from well-known celebrities have not helped the neighborhood return to its pre-disaster state due to a lack of a coordinated and holistic approach. Further, municipal planning efforts that address systemic challenges have been nonexistent, leading to mass evacuation from this once lively and culturally significant community [2].

The Lower Ninth is not unique in its struggle to pursue effective and implementable resilience strategy; however, there are examples of other disaster response efforts that have fostered an ambition to close this planning and governance gap. The Camp Fire in November 2018 destroyed the Town of Paradise and surrounding unincorporated communities in Butte County, California. Although considered one of the deadliest and most destructive wildfires in California history, residents, fire officials, and business leaders spoke of the possibility that this could once happen again, demanding a response that would better equip their town and its residents for the future. In a coordinated effort between State of Emergency and Major Disaster declarations at the state and federal level, FEMA's Long-Term Community Recovery program, and a generous donation from the Butte Strong Fund, the Town of Paradise unified under one guiding document—the "Long-Term Community Recovery Plan" (LTCRP) (Urban Design Associates, 2019). A compilation of recovery projects, the LTCRP served as both a coordinated recovery strategy and vision for mandating a resilient future in their community. Although it has been nearly eight years since the Camp Fire and the Town of Paradise has not yet returned to its pre-disaster state, recovery and rebuild efforts have been aimed at hardening the town's infrastructure, providing safe homes and streets for residents, and reconnecting the community in such a way that future wildfire response will be more coordinated, interwoven, and adaptive.

2.3 Toward a Process-Oriented, Place-Based Resilience

Recent scholarship argues for moving beyond resilience as a fixed "state" toward understanding it as an *ongoing, place-based process*. Communities continually negotiate trade-offs, balancing short-term recovery with long-term transformation. In this sense, resilience is spatially and socially embedded, shaped by local histories, governance cultures, and regional interdependencies.

This insight parallels work on energy paradigm shifts. Kelly [4] demonstrates that energy transitions unfold not only as technical substitutions but as paradigmatic shifts in social values, governance norms, and institutional arrangements. This framing underscores that resilience in the energy sector—whether through distributed generation, district energy, or waste heat recovery—must be viewed as part of a broader transformation in how societies conceptualize and manage energy.

Applied tools like the RRC Resilience Toolkit operationalize this approach, structuring resilience planning around four pillars—people, place, economy, and infrastructure. Rather than isolating risks by sector, this framework emphasizes how vulnerabilities cascade and interact, and how strategies must therefore build capacity across domains.

In a similar way, coordinated planning processes, like the Paradise Long-Term Community Recovery Plan, demonstrate the importance of collective recovery, rebuilding, and resilience planning. Bringing together multiple levels of governance, diverse resident experiences, and individual sector leaders under one process supports a co-authored future for communities where isolation and individual domains have largely been the status quo.

2.4 Infrastructure as a Frontier of Applied Resilience

Resilience theory underscores that infrastructure is not merely a technical substrate but a connective tissue that shapes community vulnerability and adaptive capacity. As Meerow and Stults (2016) argue, infrastructure planning must integrate climate mitigation, adaptation, and social equity considerations simultaneously.

Kelly-Pitou emphasizes that energy infrastructure in particular must be understood as both technical and social. Microgrids, district energy, and thermal networks exemplify how infrastructure can serve as adaptive platforms, designed to evolve with shifting conditions while reinforcing community resilience. By aligning with paradigm-shift thinking [4], such systems demonstrate resilience as a process of continuous adaptation and transformation.

Emerging research on high-performance computing (HPC) centers further illustrates this potential. When designed systemically, waste heat recovery and district energy integration transform an environmental liability into a community asset, embodying resilience as a condition of adaptive co-benefit rather than reactive recovery.

3 Operationalizing Systems Resilience: THE RRC Framework

3.1 A Holistic Approach to Community Planning

3.1.1 A Real Resilience Framework

While the broader definition of resilience is "the capacity to withstand and recover quickly from difficulties," it is important to look at community resilience through the lens of four thematic pillars: people, place, economy, and infrastructure (hereafter referred to as the resilience framework). The resilience framework is the foundation for the RRC Resilience Toolkit, produced in partnership with the Michigan Economic Development Corporation (MEDC) to provide a guide for communities with information, metrics, publicly available tools, and suggested action items that help better prepare them for lingering stressors like aging infrastructure and economic decline and abrupt shocks like flooding and wildfires. This framework also helps planners understand the interconnectedness of systems within cities to ensure planning is approached holistically.

3.1.2 Applying the Framework

Piloting the Process

One of Midland, Michigan's most defining features is its location along the Tittabawassee River, as it offers access to some of the most renowned and unique natural features in the area. However, this proximity to the river led to widespread economic and housing challenges in spring of 2020 after a heavy rain event contributed to the failure of two 95-year-old embankment dams 20 miles upriver, completely draining the Wixom and Sanford Lake reservoirs and causing major flooding in Midland [3].

The event displaced over 10,000 residents in the midst of a global health crisis; homes and businesses were destroyed, vulnerable residents were sent to shelters risking exposure to COVID-19, there was an outbreak of the disease among disaster recovery workers, and in an area already experiencing a housing crisis, the availability and affordability of housing was exacerbated by this disaster.

Rise + Resilience is a technical assistance program offered by MEDC in collaboration with SmithGroup; through this service, communities are guided through the process of using the resilience framework and the tools available in the Resilience Toolkit. Midland was one of the first communities to take part in this program, engaging with the team two years after the flooding disaster of 2020.

Over the course of a year, the city was guided through a self-assessment survey and initial focus group discussions centered around the four thematic pillars of the resilience framework, which gave insight into existing plans and processes and revealed potential opportunities for resilience initiatives. The SmithGroup team reviewed these materials, completed additional research to inform the development of a plan, and spent several months creating an extensive document with key action items across all four pillars in addition to two additional strategy memos focused on encouraging infill and strengthening neighborhood hubs.

Adapting the Process in Real Time

Northeastern Michigan experienced several rounds of freezing rain over 60 hours in early 2025 that resulted in the accumulation of between 0.25–1.0 inch of ice, sheeting cities across the region and weighing heavily on trees and powerlines [9]. This event led to widespread power outages, leaving over 100,000 people in more than a dozen counties without heat in freezing temperatures. While line workers from across the state were deployed as soon as possible, most residents were left without power for a minimum of five days, with thousands of rural community members waiting another two weeks [1] for their electricity to be restored due to inaccessibility of power poles in the densely wooded countryside.

SmithGroup was working with one of the affected communities, the city of Alpena, during this time. Prior to the storm, a newly formed resilience committee had been engaged in one workshop where members discussed challenges and concerns together after individually completing scorecards where they ranked the status of resilience indicators within categories under each pillar of the resilience framework. The workshop resulted in thoughtful discussion about the environment, housing availability, attracting new residents, and the unique culture and mindset of those who live in the area.

The topic of infrastructure was met with hesitation, as the city engineer was not present for the call. The committee discussed aging infrastructure at the cement plant in town, which is a major employer for the area, and how the city needed to request an appropriation to perform needed maintenance on critical drinking water infrastructure. The relationship between economic growth and the ability to fund infrastructure maintenance was also an important point that was made during this segment of the discussion. When asked about the stability of the energy grid, especially during major storms, committee members responded with praise for the utility company, highlighting that although power lines were above ground, the trees around them are maintained well and when outages do happen, they do not last very long.

Three weeks later, the ice storms revealed the larger region's weaknesses in energy security.

In subsequent workshops after the storms, every committee member offered valuable insight into the experience from their perspective. The systems within their city were revealed over the course of the response and recovery effort, and infrastructure became a major topic of conversation. Through these discussions, gaps in preparedness were identified, successes were documented, and thoughtful collaboration was in full effect. At the end of a five-month process, the committee in partnership with the Rise + Resilience team drafted a document which focused on two major topics, disaster preparedness and smart growth. Implementable action items related to these topics within each of the pillars of the resilience framework were chosen, and community champions were identified to ensure progress is made on each.

In contrast to how the program was deployed in Midland, the Rise + Resilience team was able to adapt the process to the needs of Alpena's community in real time and develop a collaborative experience with the committee in response to their very fresh perspectives directly following a disaster event.

3.1.3 Engaging the Community in Collaborative Planning

Forming a committee of passionate and motivated stakeholders is an invaluable first step. This can include individuals from the city (planners, engineers, local fire department, EMS), representatives from development authorities and the chamber of commerce, planners from regional- and county-level councils, representatives from major employers and small businesses, non-profit organizations, and state-level coordinators. The perspectives gained from these participants help to ensure that each pillar of the resilience framework is represented, allowing for a rounded conversation about the challenges the community faces.

Bringing this committee together in a series of workshops throughout the development of a resilience plan allows room for initial thoughts and lived experience to be shared along with feedback about how current stressors and recent shocks have been handled; it also provides a space for philosophical debate to occur. This is an opportunity for coordinated effort to work toward solutions after hearing what challenges different members of the community face and where there are opportunities for collaboration.

3.1.4 Assessing Risk and Evaluating Previous Disaster Response

Resilience actions are not universally applicable, as every city across the globe is subject to differing climatological, environmental, social, and economic conditions which expose its population to different shocks and stressors. Understanding the unique risks to which a community is exposed not only helps identify the most critical challenges needing to be addressed, but it also prioritizes fiscal responsibility in community planning efforts.

Climate projections for temperature and precipitation can provide an understanding of the changes a city may experience in the next several decades. Analyzing this information can provide indicators for potential risk, as large deviations from historical averages provide insight into the systems needing fortification. While projections are helpful in determining what future challenges affect a community, reflecting on past events and monitoring existing conditions can help in understanding and accounting for the human response to disaster events. In the case of Alpena, the recency of the ice storms allowed for fresh reflection about how the community responded, which was particularly beneficial for identifying reliable organizations for recovery efforts and opportunities for collaboration for an event that is likely to happen with greater frequency in the future. In the case of Paradise, the awareness that such a devastating wildfire could strike the town again in the future created a prioritization for hardened utility infrastructure, improved evacuation responses, and greater enforcement of fire-safe building and landscape management practices, resulting in a community that is better equipped for fire response and safer for day-to-day life. Acknowledging and anticipating the recurrence and impact of events is an important component in resilience planning, and the sooner community leaders can accept and coordinate preparedness activities and implement projects that provide effective adaptation to these conditions, the smaller the scale of damage and need for recovery resources can be.

3.1.5 The Consequences of Fragmented Planning

The events in Midland and Alpena revealed that disasters transcend city boundaries, as they are often related to shared resources and infrastructure, larger storm systems, and extend beyond the abilities of a single community to affect change.

The practice of fragmented planning can result in unnecessary expenditure of funds, as adjacent communities may unknowingly be working to solve related but separate problems. Fragmented planning can also result in unprotected assets as one community invests in housing, business development, and placemaking efforts that could be at the mercy of poor infrastructure maintenance in another city, as in the case of Midland.

Having the foresight to monitor change, analyze trends, and identify vulnerabilities facilitates the recognition of relationships between systems within a city and the larger regional context in which a city exists; this exercise can help reduce the potential for cascading failures, or when one element is pushed beyond its capacity and leads to the breakdown of other parts of the network.

3.2 *Regional Planning*

3.2.1 Scales of Influence

The events in Midland and Alpena demonstrate that, while disasters are experienced locally, their *causes* and *impacts* are often regional. This dynamic suggests that a model of coordinated resilience based on vertical integration is needed for adequate preparation, adaptation, and response—one where distinct, yet overlapping roles, are clearly defined across local, regional, state, and federal levels.

- **Local Level:** Municipalities are the front lines of resilience planning. They directly manage the land use, maintain local infrastructure, engage with local residents and stakeholders, and are the first to respond to emergencies. However, their authority and resources are limited by their jurisdiction. As Midland's story illustrates, a city can do everything "right" within their jurisdiction and still be heavily impacted by events—such as a dam failure 20 miles upstream—beyond their control.
- **Regional Level:** Entities like counties, councils of government (COGs), and planning commissions operate at the regional level and can become an intermediary coordinating body. Their ability to coordinate and manage shared systems among municipalities enables them to tackle challenges that transcend jurisdictional boundaries. They can convene stakeholders and local leaders, pool and disseminate resources, and develop strategies that address systemic risks.
- **State and Federal Levels:** These higher levels of government act as essential enablers. State agencies, like MEDC, can provide research, frameworks, funding, network building, and technical assistance (such as the RRC Resilience Toolkit) that empower local and regional action. The state itself establishes coordinated efforts and establishes the guiding policies and funding for local and regional entities. And the federal government has traditionally provided large-scale disaster aid, interstate coordination, support for large-scale resilience and infrastructure investments, and establishment of national standards.

While larger scales of government—those at state and federal levels—have broader reach and influence, their responses are often slow and complex. Regional entities can help bridge the gap between state and local entities to decrease complexity and speed up response times while still having multi-jurisdictional and systems-scale impact.

3.2.2 Coordinating Adaptation and Response Activities

Adopting regional-scale coordination requires commitment to systems thinking and multi-jurisdictional engagement. Resilience planning must move beyond localized response (like a single town's flood wall) and instead embrace integrated thinking that focuses on the health of the entire system (the watershed). This approach recognizes that the actions made in one community directly impact others within a common system.

While states often have existing councils of government which coordinate strategic investment and regional partnerships, and many of these councils have task forces that address regional resilience, they often fall short of what happens when a region-wide disaster occurs, as in the case of Alpena and its many neighboring communities. State-level response often relies on federal aid and can be a lengthy process with a messy coordination effort; an interregional approach can be an alternative solution, rooted in local suppliers and mutually beneficial partnerships.

The practical application of this approach involves several key activities:

- **Shared Risk Assessments:** Rather than each community conducting its own isolated vulnerability analysis, regional partners can collaborate on a single, comprehensive assessment of a shared resource. A framework like MEDC's Resilience Toolkit can be adapted to a regional scale to facilitate assessment and coordination. This approach creates a common, data-driven understanding of risks and allows for more strategic, coordinated, non-duplicative investments in mitigation.
- **Joint Action Plans:** Regional climate adaptation or hazard mitigation plans formalize shared goals and ensure that one community's development plans do not undermine a neighbor's resilience efforts. For example, flood control measures can sometimes inadvertently compound impacts for downstream communities. This is important for managing shared resources like waterways and areas along the Great Lakes coast.
- **Mutual Aid Agreements**: These agreements are the foundation of a coordinated response. As the multi-county power outage around Alpena demonstrated, a single community's resources can be quickly overwhelmed. Formal mutual aid agreements pre-negotiate the sharing of emergency personnel, equipment, and shelter across jurisdictions, ensuring that help can be deployed quickly and efficiently when a disaster strikes.

4 Infrastructure Integration

4.1 Applying Resilience Theory

Infrastructure is the connection that links communities to each other, allowing their residents to share natural and economic resources; roads, bridges, and lines for energy, water, and communication are just some examples of the large-scale services needed for daily life that are vulnerable to natural hazards. Because infrastructure spans across municipal borders, it can be difficult to implement effective resilience efforts without regional coordination.

Introducing small-scale solutions like thermal networks and microgrids can help reduce localized impacts associated with power outages. Working to coordinate wider, regional-scale disaster response for overlapping policy and co-beneficial investment has the potential for reducing the risk of massive destruction across multiple communities; an example of this is partnership and investment in the upgrading and maintenance of expensive infrastructure such as dams. By tackling these challenges from local and regional angles concurrently, greater resilience against multiple hazards can be established.

Understanding how infrastructure vulnerability relates to and exacerbates underlying stressors in a community's social, place-based, and economic systems can unlock a community's ability to push beyond the traditional practice of response and recovery and into the areas of mitigation and adaptation, not only to climate impacts but to technology advancements as well. As the demand for greater computing capacity increases with the ever-growing usage of cloud-based services and artificial intelligence, the more communities must be prepared for the introduction of data centers and their consumption of energy, water, and land. This presents an opportunity for a truly integrated approach that can help find a balance between the necessity of developing data centers and providing benefit by addressing some of the challenges from shocks and stressors that face a community.

4.2 Environmental Impact of HPC Operations

"One query to ChatGPT uses approximately as much electricity as could light one light bulb for about 20 minutes."—Jesse Dodge, Research Scientist, Meta [6].

It is estimated that in 2023, 4.4% of all electricity in the US was consumed by data centers [10], with experts estimating a 165% increase in global electricity demand by 2030 (50% by 2027 alone) (Goldman [11]). These numbers are staggering, and illustrate how data centers, and more specifically "High-Performance Computing Centers" (HPC's)—computing centers with focused, high-intensity operations associated with machine learning and the artificial intelligence revolution—loom large over all discussions of infrastructure, resilience, climate action and sustainability.

4.2.1 Grid and Municipal Water Burden

A single 130 MW HPC can consume as much as 200 million gallons of potable water per year, the equivalent yearly water usage of over 21 thousand homes in the US, given current data center cooling operations (See Fig. 1), all of which are required to be highly purified to minimize fouling on cooling coils and heat exchangers. At this rate, it would take less than 1900 data centers to consume as much water as New York City in a year.

Fig. 1 Water consumption for high-performance computing compared to residential consumption

4.2.2 Renewable Energy Integration

Given HPC's consistent and intensive power consumption, renewable energy sources such as wind and solar are not feasible on site to meet an HPC's needs, with even utility-scale renewable energy projects often not providing sufficient capacity. In an effort to reduce the immediate and dramatic increase in scope 2 emissions associated with the operation of these HPC's, major tech companies have brokered deals to reactivate some of the US's retired nuclear energy assets, with Microsoft announcing last year that it is working to restart Three-Mile Island (Harrisburg, PA) to power their cloud computing and artificial intelligence projects [7]. Nuclear energy is something of an ideal pair with HPC operations, due to its consistent power output and low life cycle carbon impacts. Unfortunately, given current regulatory speeds around permitting novel nuclear operating licenses in the United States, it is more common for natural gas power plants to be paired with HPC's because of their relatively inexpensive cost to develop.

These consumption figures can mean the difference between an aging coal plant being phased out by a local utility, and its necessary resurrection to deal with the massive demand for electricity generation. Finding opportunities to decrease overall utility consumption can mean the difference between "meeting the moment" as it relates to climate change and missing it completely.

4.3 Waste Heat Recovery from HPCs

One emerging frontier lies in the rapid expansion of high-performance computing (HPC) centers across the United States. While critical to the digital economy, HPCs are also energy-intensive, producing substantial amounts of waste heat that often go unutilized—thereby contributing to carbon emissions and grid stress. However, these same conditions offer an opportunity: the spatial and thermal density of HPCs makes them uniquely suited for integration with district energy systems.

To combat high-energy use intensity as well as significant water consumption that high-performance computing centers utilize in their operation, it's important to find any and all strategies that are suitable for reducing overall utility consumption. One

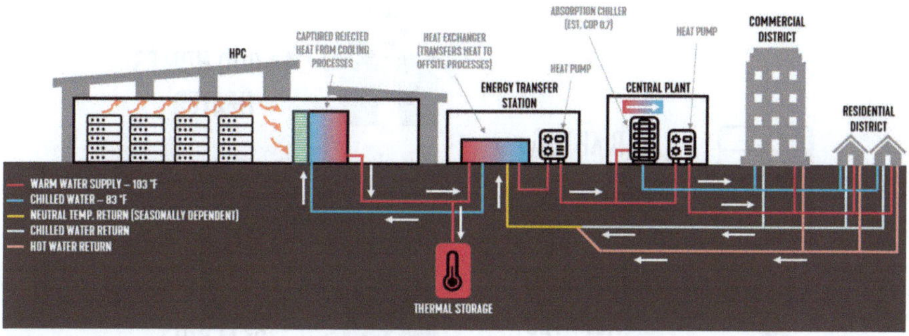

Fig. 2 High-performance computing delivering heat to a thermal energy network

highly efficient solution is the development of waste heat recovery. Waste recovery capture is made possible through a series of air-to-water heat exchangers, capturing the waste heat generated from the computing processes. This heat can then be utilized in turn as supplemental heating for not only the HPC but also for other processes on the site and even any adjacent operations. This waste heat recovery can capture heat as hot as 103 °F and with supplemental heat pumps provide excellent options for low temperature heating applications. This waste heat recovery improves site and building efficiency through elimination of supplemental space heating at the building and site level.

Rejected heat can also be integrated into larger district heating operations (See Fig. 2), which will be discussed further later in this section. This heat when supplemented and introduced to an absorption chiller can also be used to generate chilled water capacity incorporating not only heating but cooling as well, allowing for waste heat recovery to always serve as a thermal source for all climate zones at any time of the year.

In an era of increasing ecological risk due to climate change, operating these critical facilities without putting undue burden on the environment will be imperative. While high-efficiency options like distributed waste heat are utilized in other countries, the US has yet to adopt policy around means of heat capture and redistribution.

4.4 *Thermal Energy Networks*

The integration of a waste heat recovery system at high-performance computing centers, in conjunction with a large-scale thermal energy network is where efficiencies grow exponentially. This low-temperature hot water distribution at the site or even municipal scale, allows for the waste heat collected at the high-performance computing center to be utilized in the form of free or supplemented heating and cooling. For example, a 130 MW high-performance computing center can generate

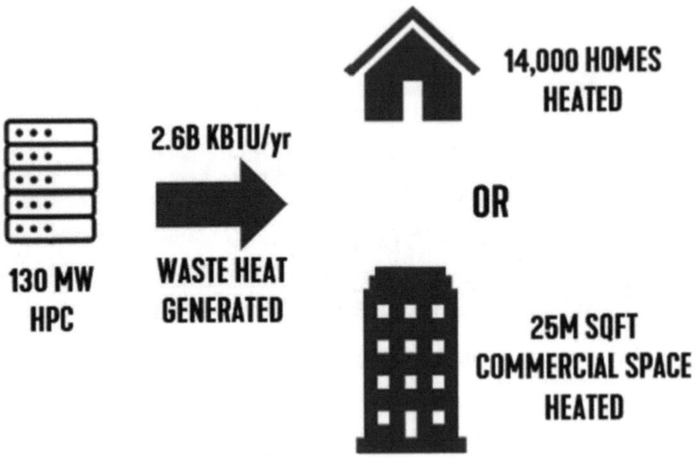

Fig. 3 Capacity comparison for high-performance computing waste heat

enough waste heat to then heat around 14,000 homes. Or for commercial and industry partners, this waste heat can be used to heat 25,000,000 sq.ft. of commercial space (See Fig. 3). These opportunities for significant improvements over traditional heating and cooling processes, in this example, steam boilers and traditional, centrifugal chillers, can see efficiencies of over seven times. In this configuration, low-temperature hot water or even ambient temperature water can be delivered to users offsite via heat exchangers at the property boundary.

4.5 District Utilization

4.5.1 Waste Heat as a Commodity

Waste heat capture is not a new technology, with waste heat recovery systems already integrated into thermal energy networks and HPC operations across Europe. Many European countries have led the way to adopting these systems, driven by ambitious sustainability and climate targets at the municipal, federal and intergovernmental levels. It is a mutually beneficial solution to a unique challenge: high-performance computing centers can repurpose the waste heat that through typical operation would have required intense mechanical cooling strategies, and can therefore reduce their overall carbon footprint. In turn, this recovered waste heat from the HPC can then serve the surrounding communities as a low-carbon utility source, reinforcing the role of HPCs as contributors to local energy resilience, ensuring redundancy and reliability.

In the United States thermal energy networks have not been widely developed as a part of broader infrastructure planning, meaning many of these systems would have to be built and financed by some form of "district heating developer." These district heating developers can be either the local utility, or third-party public or private entities. The necessary infrastructure would include the connection from the HPC to an adjacent energy transfer station that would house the heat exchangers, heat pumps, absorption chillers, etc. for the broader distribution, as well as the district loop infrastructure that would connect to the end users. Ideally, these systems would be set up around large anchor loads which can include university campuses, hospitals, science and technology campuses, or public services (community centers, libraries, museums, public recreation, social housing, etc.) which could assist in maintaining system balance.

4.5.2 Stability

The Energy Information Administration (EIA) reports that space heating is the single highest end use of energy in the United States for both commercial and residential buildings, with space heating accounting for 32% of total energy use in commercial buildings [12] and 43% of total energy use in residential buildings across all housing types (U.S. Energy Information Administration [13]). High-Performance Computing Centers have a continuous supply of waste heat available for use; as long as the servers are running, heat is being produced as a byproduct. Knowing that the HPCs are a source of constant heat, they can provide a stable foundation for the energy resilience of a local utility grid.

If absorption chillers are not used to provide cooling to a district system, the excess heat, when not needed, can charge a thermal energy storage tank or field. This stored heat would then be utilized during times of grid instability or peak utility pricing, which then helps to reduce energy costs and to support grid load balancing.

High-performance computing centers can be seen as a catalyst for decarbonization efforts here in the United States. Our research into HPC and TENs operations demonstrates that embedding energy recapture and distribution into the design of HPC infrastructure can support the development of resilient, low-carbon urban energy systems. Such innovations underscore the potential for aligning infrastructure planning with climate adaptation goals through a systemic lens. By connecting resilience theory with energy and design practice, this work offers a replicable model for how local governments, private partners, and regional agencies can co-create integrated solutions that both mitigate risk and generate long-term public value.

5 Infrastructure Planning

Infrastructure planning for resilience requires a fundamental shift from conventional approaches that treat climate adaptation, economic development, and community needs as separate planning challenges. Effective resilience planning recognizes infrastructure as the connective component that either amplifies or mitigates risks across the pillars of the resilience framework. This section outlines strategies for aligning infrastructure investments with climate goals while addressing the practical challenges of implementation and risk management.

5.1 Climate Goals

Infrastructure planning must be grounded in clear climate objectives that operate at multiple temporal and spatial scales. At the community level, these goals should address both mitigation (reducing greenhouse gas emissions) and adaptation (building capacity to withstand climate impacts). Achieving these objectives requires moving beyond project-by-project assessment toward portfolio-level thinking that considers how infrastructure investments interact across sectors and scales.

Successful climate-adapted infrastructure planning integrates several key principles:

Carbon neutrality and circularity: Infrastructure systems should be designed to eliminate waste streams and create closed-loop resource flows. The example of waste heat recovery from HPC centers illustrates this principle, transforming an energy liability into a community asset that can heat thousands of homes while reducing grid stress and carbon emissions.

Adaptive capacity building: Rather than designing buildings for static climate projections, infrastructure should enhance a community's ability to respond to changing conditions. This might include modular energy systems that can be reconfigured as demand patterns shift or green infrastructure that provides multiple co-benefits across different climate conditions.

Regional coordination: Climate impacts transcend municipal boundaries, requiring infrastructure planning that considers regional interdependencies. Energy grids, watershed systems, and transportation networks all operate at scales larger than individual communities, requiring coordinated approaches to risk management and resource sharing.

5.2 Challenges around Implementation and Managing Risk

Despite the clear rationale for climate-resilient infrastructure, significant barriers persist in translating planning frameworks into implemented projects. Understanding and addressing these challenges is essential for communities seeking to implement resilience goals.

Financial and governance complexity: Infrastructure projects often require coordination across multiple funding sources, regulatory jurisdictions, and stakeholder groups. Innovative approaches like district energy systems may involve public-private partnerships that stretch traditional municipal finance and governance models.

Communities need new frameworks for evaluating long-term value creation that accounts for co-benefits across multiple sectors.

Technical and regulatory barriers: Emerging technologies like thermal energy networks or waste heat recovery systems often encounter regulatory frameworks designed for conventional infrastructure. Communities may need to advocate for policy reforms or develop pilot project structures that demonstrate viability while working within existing constraints.

Risk assessment and uncertainty: Climate resilience requires planning for an uncertain future, which can be challenging for traditional engineering and financial approaches. Infrastructure investments must be robust across multiple climate scenarios while remaining economically viable. This demands new approaches involving risk assessment that can quantify the value of adaptive capacity and system redundancy.

Community acceptance and engagement: Infrastructure projects succeed or fail based on community support, yet resilience planning often involves technical concepts that can be difficult to communicate. Effective implementation requires ongoing community engagement that helps residents understand how infrastructure investments connect to their daily experiences and long-term security.

Timing and phasing: Building resilient infrastructure often requires coordination between multiple projects and entities across different timelines and funding cycles. Communities need strategic frameworks for phasing out investments to maximize synergies while maintaining momentum for long-term transformation.

Addressing these challenges requires moving beyond traditional project delivery methods toward more collaborative, adaptive approaches that can respond to changing conditions while maintaining focus on long-term resilience goals. The most successful infrastructure projects are those that are implemented by communities where infrastructure planning is not only approached as a technical exercise but is a collaborative process of developing community capacity that aligns immediate needs with long-term resilience objectives.

6 Implementable Actions for Community Resilience

This section can be used as a checklist to operationalize the resilience framework, allowing city staff to begin thinking about the related systems that exist in their communities.

6.1 Committee Formation & Community Workshopping

Identify community stakeholders within each of the thematic areas of the resilience framework and invite them to take part in a series of resilience workshops. Each workshop can be centered around a single theme, focused on discovering ongoing

Table 1 Suggested community workshop participants

People	Place	Infrastructure	Economy
· Fire department. · City police/sheriff. · Hospital admin. · Emergency management. · Social work orgs. · Local churches. · Food pantries. · Housing developers. · Local government. · Non-profit orgs. · Neighborhood groups.	· Parks & rec. · State historic preservation office. · Dept. natural resources. · Agricultural groups. · Environmental groups. · Nature conservation advocacy groups. · Tribal reservations.	· City engineering. · Waste water treatment. · Waste management. · Energy service providers. · Dept. of transportation. · ADA advocacy groups. · Ham radio clubs. · Line workers.	· Local business owners. · Major employers. · Commercial developers. · Adult services. · Childcare orgs. · Local colleges. · Public schools. · Investors. · Philanthropists. · Chamber of commerce. · Visitors bureau.

challenges and those that each stakeholder encounters prior to, during, and after a disaster. These discussions can help to identify common and recurring issues and provide an opportunity for stakeholders to make connections that may lead to future partnerships and collaboration.

Leaders or representatives from the following groups can provide valuable insight during community workshops (See Table 1):

During the workshops, identify stakeholders who are interested in participating as members of a committee and representing the groups involved in the discussions of their resilience pillar. The members of this committee will be responsible for not only co-creation of the community resilience plan, but also for advocating for the groups they represent, identifying champions, and activating the action items within the plan.

6.2 Risk Perception

This step involves understanding what the committee, with community input, sees as a risk. Compiling the information shared in the community workshops into one space can help the committee to evaluate which challenges are shared between the pillars and to identify the shocks and stressors to which they are most vulnerable across each theme. For Rise + Resilience, the team created a scorecard to help structure these discussions and identify which areas require the most attention (See Fig. 4).

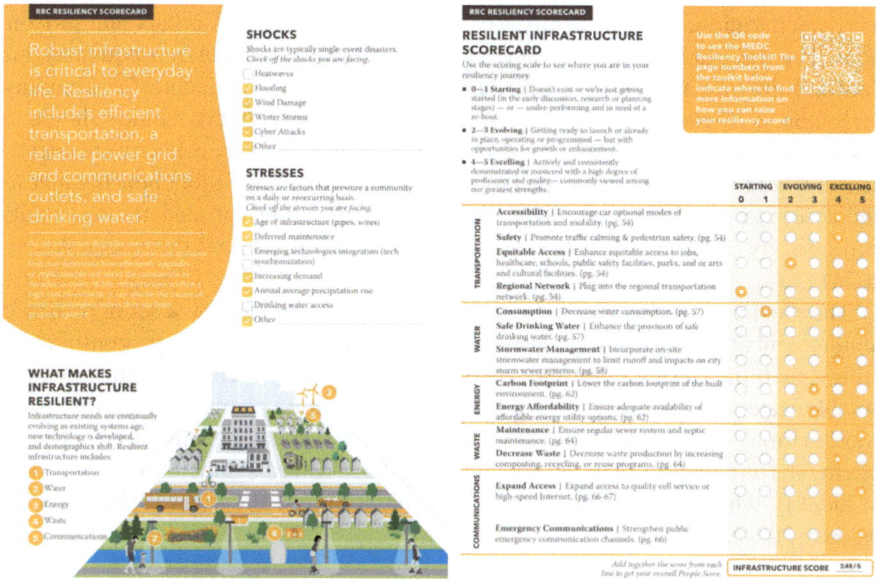

Fig. 4 MEDC RRC Resilience Scorecard

6.3 *Risk Evaluation*

Risk evaluation translates perceived risks and lived experience into a structured basis for decision-making. While risk perception captures what stakeholders believe to be most threatening, evaluation applies consistent criteria across shock types and across the four pillars of the resilience framework—people, place, infrastructure, and economy—to produce a prioritized set of scenarios defined by likelihood, severity, duration, and cascading potential. This approach aligns with resilience literature that frames urban vulnerability as an outcome of infrastructure interdependence and governance conditions rather than isolated asset performance.

A systems approach evaluates risk as interactions between hazards and interconnected services with identifiable disruption pathways. Extreme events rarely affect a single sector; initial failures can propagate into secondary and tertiary impacts across energy, water, transportation, and communications, amplifying consequences and complicating recovery. Scenario-based framing makes these interdependencies explicit—for example, an ice storm can trigger grid failure, reduced heating access, disrupted communications, constrained transportation, and strained emergency response capacity. Flood risks are similarly evaluated most effectively at watershed or regional scales where upstream conditions, shared infrastructure, and multi-jurisdictional governance shape local outcomes.

To support transparent prioritization, municipalities can use a rubric that assesses probability within a planning horizon, consequence to critical services, expected recovery duration, and—critically—cascading potential across pillars and jurisdictions. Federal guidance also emphasizes incorporating indirect consequences and known vulnerabilities, since direct damage is only one component of community-scale impact (Federal Emergency Management Agency 2025). Equity should be embedded as a core criterion because risk exposure and recovery capacity are uneven; heat and outage impacts are shaped by affordability, baseline health, housing quality, and access to social supports, with vulnerability often concentrated among populations already experiencing economic strain (U.S. Environmental Protection Agency 2024).

The most useful outcome is not only a ranked risk list but also a set of resilience requirements that guide mitigation and infrastructure investment (e.g., continuity of critical facilities during multi-day outages, thermal safety during extremes, reduced potable water vulnerability during peak stress, and strengthened interjurisdictional coordination). Spatial and infrastructure data tools can support this translation by pairing hazard exposure with social and infrastructural vulnerability indicators to identify where compounding risks concentrate (U.S. Environmental Protection Agency 2024).

6.4 Risk Mitigation

Risk mitigation translates evaluated priorities into coordinated strategies that reduce vulnerability and strengthen adaptive capacity across the pillars of people, place, infrastructure, and economy. While traditional hazard planning emphasizes protecting individual assets, systems-based resilience focuses on reducing the likelihood and severity of cascading disruptions and improving a community's ability to absorb shocks and reorganize under stress. This shift requires moving beyond isolated projects toward integrated interventions that combine governance improvements, operational readiness, and infrastructure investment. Research on infrastructure interdependencies demonstrates that mitigation must address both physical vulnerabilities and the institutional and coordination conditions that determine whether disruptions remain localized or escalate into system-wide failure.

Mitigation operates across multiple, complementary levels. Some actions reduce exposure through land use and development policies that limit growth in high-risk areas and strengthen protective landscape functions. Others increase robustness and continuity through redundancy measures such as backup power for critical facilities, diversified communications pathways, upgraded water and wastewater systems, and improved emergency access routes. Governance-oriented actions are equally critical, including staff training, procurement readiness, interjurisdictional agreements, and coordination structures that reduce response friction during compound events. Many communities experience disasters as coordination failures as much as engineering failures, particularly where responsibilities and resources are distributed across local, regional, and state actors (Federal Emergency Management Agency 2025).

A core principle of systemic mitigation is evaluating actions not only for immediate protection but also for their ability to generate cross-pillar co-benefits. Prioritizing strategies that simultaneously support public health, economic stability, and infrastructure reliability reduces fragmentation and strengthens overall resilience. Equity is central at this stage, as mitigation investments can unintentionally reproduce vulnerability when benefits accrue primarily to already-resourced areas. Research on extreme heat demonstrates that financial barriers, health vulnerabilities, housing quality, and access to cooling shape risk outcomes, underscoring the need to consider distributional impacts, community participation, and programs that reduce energy burden and improve thermal safety during extreme events.

Mitigation also includes transformational strategies that treat infrastructure as an adaptive platform rather than a static system. In the energy sector, the convergence of climate-driven disruption and rising electricity demand makes energy planning a central resilience concern, as grid instability and utility burden amplify risks across all pillars. Thermal energy networks offer a scalable approach to improving heating and cooling efficiency while integrating multiple thermal sources, including data centers, wastewater, and ambient geothermal exchange (International Renewable Energy Agency n.d.). Research on data center waste heat recovery highlights growing global attention to these systems, particularly where stable thermal output and proximate demand align. When paired with district-scale distribution and storage, waste heat recovery from high-performance computing centers can reduce resource intensity during normal operations while supporting thermal continuity during disruptions.

The mitigation phase culminates in a prioritized project portfolio and an implementation pathway that reflects municipal capacity, partnership requirements, and phasing. Because many resilience interventions involve interdependent systems and regional infrastructure, mitigation planning must identify enabling governance structures and delivery models, including roles for utilities and third-party developers. Applied guidance increasingly emphasizes resilience planning tools that support cross-sector decision-making and translate risk priorities into implementable investments (Federal Emergency Management Agency 2025). Framed as a coordinated systems agenda, infrastructure innovations such as thermal energy networks and waste heat recovery function not as standalone sustainability projects, but as resilience interventions that address cascading risk, equity vulnerabilities, and long-term climate and development objectives simultaneously.

Defining Resilience uniquely to community vulnerabilities While resilience has a common definition, it is important to ensure communities have a shared understanding of what resilience means to them. This allows the unique perspectives of community members and the place to be represented in its vision for the future. By putting resilience into a shared language that is backed by the information gathered in the community workshops, the word becomes personal and applicable to the people who will be doing the work.

For Alpena, resilience means "involving the community, and building a strong foundation through preparation and structure; knowing how to respond to adversity, and doing what you have to do to take care of everyone; knowing who you can count on and what outside resources are available to make it through; and being able to adapt to change, move forward, and make your community better than it was before."

6.4.1 Plan Audit

Conducting an audit of previous planning efforts can help to understand and build upon the community's existing efforts and priorities. This is a good opportunity to identify gaps in those plans that may have been discussed in the community workshops, to highlight ongoing challenges that have not been addressed, and to find actions related to community resilience that have stalled in the years since the existing plans have been adopted. This is also an opportunity to zoom out from the community scale and to understand the city's planning efforts in relation to regional and statewide plans; this exercise can help in finding partnership opportunities for shared challenges.

Plans to evaluate can include the community's comprehensive plan, parks and recreation plans, emergency management or hazard mitigation plans, transportation plans, county-wide efforts, neighboring community plans, and local and statewide initiatives that may be related to the resilience framework.

6.4.2 Project Identification

Project identification translates evaluated risks and resilience requirements into a prioritized set of implementable actions, moving from diagnosis to investment. This phase is critical because resilience plans often fail not due to a lack of vision, but because actions are poorly scoped, sequenced, or mismatched to local capacity. Project identification should therefore be a structured decision-making process aimed at assembling a portfolio that reduces Tier 1 risks, interrupts cascading failures across the four pillars, and delivers co-benefits during non-disaster periods.

A systems-based approach classifies candidate actions by how they intervene along disruption pathways: reducing exposure, increasing robustness and redundancy, improving coordination and response capacity, or transforming systems to deliver ongoing co-benefits. Flood risks, for example, may yield stormwater, land use, and watershed coordination projects, while outage risks may yield backup power, emergency sheltering, communications redundancy, and continuity of operations planning. This typology discourages single-sector solutions and supports multi-pillar outcomes while clarifying governance roles and delivery partners.

Projects should be screened using criteria that balance risk-reduction performance and implementation feasibility, including reduction in likelihood or consequence, mitigation of cascading impacts, and benefits to critical services and vulnerable populations. Equity should be embedded by assessing whether projects reduce barriers related to energy burden, housing insecurity, mobility, and access to heating and cooling, with guidance emphasizing participation and opportunity outcomes in addition to aggregate benefit (International Council for Local Environmental Initiatives

2024). Early consideration of cost-effectiveness improves fundability and alignment with federal mitigation requirements (Federal Emergency Management Agency n.d.).

Project identification is most effective when treated as a portfolio that distinguishes between near-term readiness actions—such as mutual aid agreements, continuity planning, and emergency communications—and longer-term capital investments, including facility hardening, backup generation, water system upgrades, and transportation resilience. Sequencing these actions maintains momentum while building capacity for transformative investments.

Within this portfolio, energy and thermal resilience strategies are particularly relevant where outages, extreme temperatures, and utility burden are priority risks. Thermal energy networks improve heating and cooling efficiency while enabling distributed supply, redundancy, and integration of diverse thermal sources, including wastewater, ambient geothermal exchange, and waste heat from data centers (International Renewable Energy Agency, n.d.). Municipal coordination with utilities around siting, rights-of-way, and phased expansion is often required.

Waste heat recovery from high-performance computing centers fits this framework as a mitigation strategy with resilience co-benefits. When integrated into thermal networks, waste heat recovery can reduce conventional heating loads, lower emissions, and provide stable thermal supply during grid stress, with feasibility shaped by demand density, temperature requirements, and governance and financing arrangements.

The outcome of project identification should be a prioritized list with clear ownership, preliminary scope, and defined next steps, ensuring projects are implementable, fundable, and aligned with community-defined resilience objectives (Federal Emergency Management Agency n.d.).

7 Conclusion

Resilience in the twenty-first century is not the simple recovery of isolated systems after discrete disruptions. It is a systemic condition that emerges from the interaction of people, place, economy, and infrastructure. The Redevelopment Ready Communities (RRC) Resilience Toolkit shows how this idea can be put into practice. It gives local governments a usable framework to anticipate risks, set priorities, and align strategies across sectors. The experiences of Midland and Alpena show that durable resilience comes from collaboration, learning, and adaptation at municipal and regional scales, not from technical measures alone.

These Michigan cases also show that shocks rarely stay within one jurisdiction or one sector. Floods, outages, and economic disruptions spread across governance, infrastructure, and social networks. This reality calls for coordination that is vertical across local, state, and federal levels, and horizontal across neighboring jurisdictions and systems. Regional planning that uses shared risk assessments, mutual aid agreements, and joint action plans is essential for reducing cascading vulnerabilities and supporting equitable recovery.

Resilience also advances through infrastructure strategies that turn liabilities into assets. High-performance computing centers offer one example. When viewed through a systems lens, their steady stream of low-grade heat can anchor thermal energy networks. These networks can lower carbon emissions and strengthen community resilience at the same time. This approach reflects a broader shift in energy and infrastructure planning. Systems are moving from linear and siloed to integrated, adaptive, and circular, with benefits that appear across multiple scales.

Significant barriers remain. Financial complexity, regulatory misalignment, technological uncertainty, and public acceptance can slow progress. These barriers point to the importance of collaborative governance, clear and transparent decisions, and sustained community engagement. Success depends on the performance of infrastructure, but it also depends on the capacity of communities to weigh trade-offs, try new ideas, and sustain collective action over time.

By connecting theory, planning practice, and infrastructure design, this work offers a model that others can use. Municipalities, regional partners, and private actors can build integrated solutions that reduce risk, advance climate goals, and create long-term public value. Resilience, understood systemically, is not an endpoint. It is a continuing process of anticipation, absorption, adaptation, and transformation.

References

1. Brewer, R (2025, April) *Thousands in northern Michigan still without power after historic storm.* Retrieved from WGVU News: https://www.wgvunews.org/news/2025-04-15/thousands-in-northern-michigan-still-without-power-after-historic-storm
2. Hamel, J (2019) Closing the gap: tiny homes, social capital, implementation, and disaster recovery. University of Oklahoma. Retrieved from https://shareok.org/server/api/core/bitstreams/0db99528-8af5-4df2-b172-cdde2d4572c9/content
3. Hayes, J (2020, August) The 2020 Midland County Dam Failure. Retrieved from https://www.mackinac.org/2020_midland_dam_failure
4. Kelly, K (2017) Paradigm shifts in energy: a systems perspective on transitions and resilience. Doctoral Dissertation, University of Pittsburgh. doi:https://doi.org/10.13140/RG.2.2.24762.03525
5. Kelly-Pitou, K (2017) Microgrids and resilience: using a systems approach to achieve climate adaptation and mitigation goals. Electrcity J, *Volume 30*(Issue 10), 23–31. doi:https://doi.org/10.1016/j.tej.2017.11.008
6. Kerr, D (2024, July 12) *AI brings soaring emissions for Google and Microsoft, a major contributor to climate change.* Retrieved October 2, 2025, from NPR: https://www.npr.org/2024/07/12/g-s1-9545/ai-brings-soaring-emissions-for-google-and-microsoft-a-major-contributor-to-climate-change
7. Mandler, C (2024, September 20) *Three Mile Island nuclear plant will reopen to power Microsoft data centers.* Retrieved from NPR: https://www.npr.org/2024/09/20/nx-s1-5120581/three-mile-island-nuclear-power-plant-microsoft-ai
8. NOAA (n.d.) *Billion-Dollar Weather and Climate Disasters.* Retrieved from National Centers for Environmental Information: https://www.ncei.noaa.gov/access/billions/
9. NOAA NWS Gaylord, M (2025) *March Goes Out Like a Lion: 2025 Northern Michigan Ice Storm.* Retrieved from https://storymaps.arcgis.com/stories/7ef8b40a604b4550beb0c3bb0658e2f8

10. Reed, R (2025, September 3) How data centers may lead to higher electrcity bills. Retrieved from Harvard Law Today: https://hls.harvard.edu/today/how-data-centers-may-lead-to-higher-electricity-bills/
11. Goldman, Sachs (2025, February 4) AI to drive 165% increase in data center power demand by 2030. Retrieved from Goldman Sachs: https://www.goldmansachs.com/insights/articles/ai-to-drive-165-increase-in-data-center-power-demand-by-2030
12. U.S. Energy Information Administration (2022, December) *Use of energy explained - Energy use in commercial buildings.* Retrieved from U.S. Energy Information Administration: https://www.eia.gov/energyexplained/use-of-energy/commercial-buildings.php
13. U.S. Energy Information Administration (2015) Use of energy explained - energy use in homes. Retrieved from U.S. Energy Information Administration: https://www.eia.gov/energyexplained/use-of-energy/homes.php

Designing Resilience in Practice: Lessons, Transitions, and the Future of Climate-Responsive Cities

Ali Cheshmehzangi, Jian Zuo, Ayyoob Sharifi, Rongpeng Zhang, Abbas Ziafati Bafarasat, and Jie Zhao

1 Reframing Climate Resilience: Key Insights from the Book

Together, the chapters in this collection show that a single design method, technological advancement, or planning tool cannot adequately address climate resilience in the built environment. Rather, resilience results from the interplay of institutional coordination, technology systems, environmental understanding, and spatial design. This book's main contention has been that climate-responsive cities and structures necessitate an integrated strategy that blends systemic innovation and architectural intelligence. This dual viewpoint, i.e., design intelligence and systemic pathways, has guided the investigation of climate-resilient built environments across the chapters.

A. Cheshmehzangi (✉) · A. Sharifi
School of Architecture, Design and Planning (ADP), The University of Queensland, Brisbane, Australia
e-mail: A.Chesh@uq.edu.au

J. Zuo
School of Civil Engineering and Construction Management, Adelaide University, Adelaide, Australia

A. Sharifi
The IDEC Institute and Network for Education and Research On Peace and Sustainability (NERPS), Hiroshima University, Hiroshima, Japan

A. Cheshmehzangi · R. Zhang
School of Architecture and Planning, Hunan University, Changsha, China

A. Ziafati Bafarasat
School of the Built Environment, Oxford Brookes University, Oxford, UK

J. Zhao
The Weitzman School of Design, The University of Pennsylvania, Philadelphia, USA

P3Global, Dubai, UAE

A. Cheshmehzangi et al. (eds.), *Design Innovation for Climate-Resilient Cities and Built Environments*, Urban Sustainability, https://doi.org/10.1007/978-981-92-2297-1_12

The ability of the built environment to adapt to climate conditions through form, orientation, materiality, and spatial configuration is the first step toward resilience at the architectural and urban design level. The conversations in Part I demonstrate how deeply ingrained climate-responsive design is in historical and vernacular architectural traditions rather than being a completely novel idea. Architecture has traditionally acted as a mediator between human settlements and environmental circumstances, as evidenced by strategies developed throughout millennia, such as passive ventilation systems, water management infrastructures, climate-responsive geometries, and adaptive urban morphologies. These examples show that rediscovering and reinterpreting spatial practices that were historically created to deal with harsh climatic situations might be beneficial for many current climate concerns.

The case studies in this book, however, also demonstrate how modern climate threats function inside far more intricate metropolitan systems. Cities are becoming extremely dynamic socio-technical environments due to rapid urbanization, increased density, energy demand, and infrastructure dependency. Therefore, the degree of resilience needed for future urban development cannot be provided by architectural design alone. For climate-responsive settings to operate well at various scales, design must be supported by digital tools, technological infrastructures, and systematic planning techniques.

This understanding has influenced the book's second major contribution, which is the incorporation of innovative infrastructure and technology into the conversation about climate resilience. The potential for how cities and buildings interact with their surroundings has increased thanks to developments in digital modelling, artificial intelligence, smart building systems, and climate-responsive materials. With the use of these technologies, planners, engineers, and designers may model climatic scenarios, track environmental performance, and optimize infrastructure and spatial solutions prior to implementation. In this way, the built environment increasingly functions as an adaptive system that can detect, react to, and learn from environmental feedback.

However, resilient results are not guaranteed by technological capabilities alone. Part II's chapters highlight the need for technology innovation to function within more comprehensive planning frameworks, governance structures, and sector-wide adoption paths. Urban landscapes, transportation networks, water infrastructure, and energy systems are all intricately linked, and the effects of climate change rarely affect any one system separately. Therefore, coordinated initiatives that connect building-level innovations with city-scale infrastructure and long-term planning objectives are essential for effective climate resilience. Taken together, the contributions in this book show how multidisciplinary climate resilience is at its core. Planning, engineering, architecture, urban design, and technological advancement must all function as complimentary parts of a broader urban system. Cities are able to transition from reactive responses to climate threats to proactive tactics that anticipate environmental change and include adaptive capacity into the built environment itself thanks to the connection between these fields.

This viewpoint reframes resilience as a dynamic design and governance process rather than a defensive reaction to climate concerns. It is necessary to think of buildings and urban systems as dynamic entities that can adjust to societal change, technology advancements, and environmental stresses. Through the integration of technological innovation, ecological knowledge, and spatial intelligence, the built environment may take an active role in mitigating climate issues.

This comprehensive understanding of climate-responsive urban development is furthered by the essays included in this edition. They show that the capacity to integrate these components into cohesive urban systems will be just as important to the future of climate-resilient cities as creative design concepts or cutting-edge technologies. The last thoughts that follow, which look at new trends, lessons from the case studies, and potential future paths for study and practice in climate-resilient architecture and urban development, are based on this synthesis.

2 Emerging Trends, Persistent Challenges, and Expanding Possibilities

The priorities of design, planning, building, and urban infrastructure are changing due to climate change, which is causing a significant upheaval in the built environment sector. The growing focus on climate-responsive design, especially through passive environmental methods and nature-based solutions, is one of the most noticeable trends. To lessen reliance on energy and enhance environmental performance, architects and planners are increasingly rethinking climatic design ideas, such as natural ventilation, shading systems, water-sensitive landscapes, and thermally responsive urban form. Concurrently, the way that structures and urban systems interact with natural processes is being influenced by ecological design techniques and biomimetic thinking.

The potential for climate-responsive landscapes is likewise being quickly expanded by technological innovation. Designers and engineers can now more accurately assess building performance, forecast climatic risks, and optimize urban systems thanks to developments in artificial intelligence, environmental sensing, digital modelling, and simulation technologies. Smart infrastructure networks, climate-responsive materials, and integrated energy systems are beginning to reshape how cities manage resources, regulate environmental conditions, and adapt to changing climate patterns.

Despite these encouraging advancements, several structural issues still impede further development. Design, infrastructure planning, and technology deployment are frequently divided into disparate sectors by fragmented governance systems. Innovative solutions' scalability is further constrained by uneven technology adoption across industries and geographical areas. Financial limitations, legal restrictions, and the ongoing disconnect between design aspirations and practical execution continue to be significant challenges. In the future, connecting architectural innovation with technological systems and planning frameworks will be crucial to the success of climate resilience measures. Transforming emergent design and technological advancements into significant changes in the built environment will require scalable infrastructure, support legislative frameworks, and coordinate sector uptake.

3 Lessons from the Case Studies: Design, Technology, and System Integration

The case studies in this collection offer verifiable proof that the convergence of environmental responsiveness, systemic technological assistance, and spatial design intelligence is necessary for climate resilience in the built environment. Several crucial lessons about how infrastructure, architecture, and urban systems may all work together to adapt to climate challenges are revealed throughout the chapters.

The first lesson is on the lasting importance of passive climate intelligence. The study of Yazd's badgirs and qanats shows how ventilation towers and subterranean water channels work as integrated environmental systems that can control temperature and support urban life in extremely dry conditions. In a similar vein, the geometry-led residential design investigated in Lahore demonstrates how spatial configuration, including courtyard layouts, building orientation, and volumetric modulation, can considerably lower thermal stress without largely depending on mechanical systems. These examples demonstrate that urban shape and climate-responsive form are still effective adaptation strategies.

The significance of ecological and biomimetic thinking in modern design innovation is emphasized in a second course. The Aqua Vascular Tower project serves as an example of how biological concepts can influence vertical architecture that can control water flows, microclimates, and the environment in densely populated areas. Simultaneously, adaptive reuse can improve environmental performance while maintaining urban memory, lowering the environmental cost of demolition and reconstruction, as demonstrated by the conversion of industrial heritage sites into climate-resilient cultural venues.

Examining housing choices in various worldwide contexts further demonstrates the need for climate resilience to address both socioeconomic and climatic realities. Effective solutions require striking a balance between environmental design techniques, affordability, cultural practices, and local climate circumstances, as demonstrated by comparative housing options.

Another critical enabling aspect is technological innovation. AI-enabled building benchmarking enables better informed decision-making throughout design and operation by offering tools for assessing building performance across energy, comfort, and health criteria. Transportation infrastructure can be converted into energy-producing urban systems through infrastructure-scale innovation, such as road pavement solar collectors combined with source-sink theory.

Finally, the case studies emphasize the crucial role of systemic integration. Resilient built environments necessitate coordinated sector adoption, supportive regulatory frameworks, and planning systems that can match building-level innovations with city-scale infrastructure networks, according to strategies for green building technology adoption and systemic planning approaches. When taken as a whole, these lessons highlight the necessity of climate resilience operating concurrently at the levels of urban government, technology, and design.

3.1 Knowledge Gaps and Future Research Directions

The case studies presented throughout this book reveal both the progress made in advancing climate-resilient design and the critical knowledge gaps that remain across the built environment sector. The combination of digital technologies with design intelligence is one of the most obvious gaps. Although environmental performance is increasingly assessed using AI-driven modelling, performance benchmarking, and simulation tools, their methodical integration into architectural and urban design workflows is still inconsistent. Therefore, future studies must investigate how digital tools might become integrated into design decision-making processes rather than only providing analytical help.

Climate-responsive housing in rapidly urbanizing areas is another subject that needs further research. Numerous examples show that passive design techniques and spatial arrangements can greatly enhance climatic performance; nonetheless, it is still difficult to use these techniques in large-scale housing delivery. In rapidly expanding cities where housing demand is rising, research must focus on how affordability, density, and environmental performance can coexist.

Additionally, there is still little long-term performance review. The environmental, social, and operational effects of many climate-resilient design techniques are not well tracked over time. Understanding how resilient buildings and urban systems function under changing climate circumstances will require systematic post-occupancy evaluation and long-term research. Lastly, future studies need to focus more intently on institutional frameworks and governance. In addition to design innovation, planning systems that can facilitate coordinated adaptation across infrastructure networks, regulatory frameworks, and urban development strategies are essential for climate resilience. The impact of emerging research areas, such as AI-assisted design, urban digital twins, climate-adaptive materials, and multi-scalar modelling, will rely on interdisciplinary cooperation between environmental science, architecture, urban systems engineering, and policy innovation.

4 Toward the Next Generation of Climate-Resilient Cities

Climate resilience is no longer a specialized issue that should only be addressed when environmental hazards become apparent. Rather, it needs to be integrated into the regular processes of design, building, infrastructure development, architecture, and urban governance. The built environment needs to change from being a reactive system that reacts to environmental shocks to a proactive system that can foresee climate stresses and gradually adapt.

Both institutional practice and design culture must change to make this transition. Using ecological design principles, adaptable urban morphology, and passive environmental principles, architects and urban designers must prioritize climate-responsive spatial strategies. The monitoring, optimization, and long-term performance of buildings and urban systems must be supported by technology systems, from digital modelling platforms to smart infrastructure networks.

The roles of cooperation and governance are equally significant. Coordination between architects, planners, engineers, legislators, and communities is necessary for climate-resilient urban development. To enable equitable and scalable climate adaptation, infrastructure investments, planning policies, and regulatory frameworks must all be in line with design innovation. In the end, how cities deal with climate uncertainty is largely determined by the built environment sector. The next generation of cities, those that are resilient, adaptable, and able to maintain human and environmental well-being in a rapidly changing climate, can be achieved through integrated design, technological innovation, and systemic urban thinking.

Index

A
Adaptive reuse, 4, 112, 121–129, 131–137, 254
Adaptive strategies, 63
AI-driven, 179, 182, 187, 255
Analytic Hierarchy Process (AHP), 58, 63–67, 69, 70

B
Badgirs, 4, 58–64, 69, 254
Biomimicry, 96–99, 109, 113–115, 118
Building performance, 91, 177–183, 186, 253, 254
Built environment innovation, 256

C
Cardiovascular-inspired design, 95
Climate adaptation, 11, 14, 17, 34–36, 41, 57, 59, 69, 76, 91, 92, 161, 163, 207, 226, 234, 239, 256
Climate adaptation strategies, 147
Climate change, 2, 4, 5, 11, 14, 39, 43, 57, 69, 76, 96, 119, 121, 141–145, 148–150, 152, 155, 156, 165, 177, 186, 187, 193, 206, 211, 212, 221, 225, 226, 236, 237, 252, 253
Climate resilience, 2–6, 11–14, 16, 35, 40, 41, 65, 68, 69, 81, 83, 110, 119, 122, 123, 126, 136, 137, 141, 150, 162, 178, 180, 194, 197, 208, 241, 251–255
Climate-resilient cities, 2, 5, 12, 13, 17, 24, 37, 40, 43, 206, 211, 253, 255
Climate-responsive architecture, 2, 5, 127
Climate risk, 11, 95, 150, 152, 157, 165, 188, 208
Computational Fluid Dynamics (CFD), 105, 119, 206–208
Construction sector, 211, 212
Cultural spaces, 121

D
Design innovation, 6, 150, 254–256
Design intelligence, 2–4, 6, 251, 254, 255

E
Energy benchmarking, 177, 179–181
Energy resilience, 57–61, 63–65, 67, 69, 75, 76, 81, 83, 88, 91, 92, 238, 239

G
Geometry-led design, 83, 88, 89
Global North, 141, 142, 144, 145, 147–150, 152, 155–165, 167
Global South, 69, 92, 141, 142, 144, 145, 147–151, 153, 155–165, 167, 168
Green building technologies, 211, 255

H
High-Performance Building (HPB), 179, 186–188
Hot semi-arid cities, 76, 88, 90
Housing resilience, 142, 144, 149–153, 155, 156, 162, 165, 167, 168

A. Cheshmehzangi et al. (eds.), *Design Innovation for Climate-Resilient Cities and Built Environments*, Urban Sustainability, https://doi.org/10.1007/978-981-92-2297-1

I
Indoor Environmental Quality (IEQ), 12, 177, 179–184, 187
Industrial heritage, 122–126, 128, 132, 136, 137, 254

L
Low-carbon architecture, 129

N
Nature-based design, 12, 13, 17
Nature-Based Solutions (NBS), 12, 24, 29–34, 39, 40, 59

O
Occupant health, 179, 181, 182, 187

P
Passive cooling, 6, 60, 61, 64, 76, 77, 80, 81, 97, 106–109, 115–118, 149, 152, 157, 158, 207
Passive design, 11–14, 16–19, 22–26, 34–38, 40–43, 59, 76, 124, 130, 149, 150, 152, 155, 160, 164, 177, 255
Potential strategies, 211

Q
Qanats, 4, 58–61, 63–66, 69–71, 154, 254

R
Rainwater harvesting, 21, 27, 96, 106, 107, 109, 111, 116, 117
Regional planning, 233, 247
Renewable energy, 78, 126, 157, 164, 194, 205, 207, 208, 236, 245, 247
Residential architecture, 4
Resilience, 2–5, 11, 12, 21, 24, 29, 32, 39, 58, 59, 63–65, 69–71, 76, 78, 80, 83, 85, 89–92, 96–98, 113, 117, 118, 121–123, 126, 128, 133, 135, 137, 142–145, 147–149, 152, 155, 156, 167, 178, 186–188, 207, 225–235, 239–248, 251, 252
Resilience framework, 65, 226, 229–231, 240, 241, 243, 246
Risk mitigation, 244
Road Pavement Solar Collectors (RPSC), 197–199, 201, 202, 204–208, 254

S
Smart building technologies, viii
Social-Ecological-Technological Systems (SETS), 58–60, 63–67, 69–71, 76, 85, 87, 134, 144, 156, 201, 204, 239, 243, 244, 246, 247
Source–sink theory, 208
Structural Equation Modeling analysis, 213
Sustainable urban development, 141, 212
Systematic Literature Review, 12, 13, 40, 65
Systems approach, 243

T
Thermal Comfort, 18, 25, 40, 41, 91, 92, 97, 108, 109, 132, 147, 149, 151–153, 157, 177, 180–186, 194, 206

U
Urban development, 2, 5, 6, 29, 34, 38, 42, 43, 77, 102, 115, 122, 150, 194, 252, 253, 255, 256
Urban Heat Island (UHI), 11, 13, 14, 17, 18, 22, 40, 42, 76, 86, 144, 149, 193–197, 201, 203–207
Urban infrastructure systems, 6
Urban regeneration, 122–124, 126
Urban sustainability, 14, 34, 35

V
Vernacular Architecture, 3, 65, 145, 147, 156, 165
Vertical architecture, 4, 96, 118, 254

Y
Yazd, 4, 57–60, 62–65, 69–71, 153, 154, 254

MIX
Papier aus verantwortungsvollen Quellen
Paper from responsible sources
FSC® C105338

If you have any concerns about our products,
you can contact us on
ProductSafety@springernature.com

In case Publisher is established outside the EU,
the EU authorized representative is:
Springer Nature Customer Service Center GmbH
Europaplatz 3, 69115 Heidelberg, Germany

Printed by Libri Plureos GmbH
in Hamburg, Germany